PRA

On Fa
Rural Communities

T0102156

"Apps has a knack for mixing facts based on data with lifelong rural experiences to spin a compelling story about life in rural America. I highly recommend *On Farms and Rural Communities*, as it continues this tradition to help us understand changes in the rural economy and the interesting impacts on rural communities."

—**Andy Lewis**, Professor Emeritus,
University of Wisconsin-Extension

"Apps's most forceful and impassioned book . . . calls our attention to the present state of rural communities in relation to the future health of our country. . . . With his passionate love of rural communities, Jerry Apps has penned a cautionary tale about what we will lose if we continue to let them diminish and the impact that will have on the rest of society. It is essential reading for both urbanites and rural residents to understand what is at stake."

—**Philip Hasheider**,
farmer, writer, and local historian

"Much has been said and written about the growing divide between those who populate our cities and suburbs and those who inhabit our rural areas. This book-length essay on what is happening in rural America and how it came to be is essential reading for those who wish to understand what is at stake. . . . The issues [Apps] raises are at the heart of what a system of rural economic health might look like in a culture that values sustainability for the people and the environment."

—**Dennis Boyer**, Fellow Emeritus,
Interactivity Foundation and Director
of Policy Projects on Agriculture and Rural Life

"If you care about rural America, read this book. If you don't care about rural America, read it anyway. This is a love note to rural farming but also a cautionary tale to all of us about what happens when we orient our economy around greed rather than well-being. It is also vintage Jerry Apps, so you'll get a heavy dose of hope and possibilities as well."

—**Katherine Cramer**,
Natalie C. Holton Chair of Letters & Science,
Virginia Sapiro Professor of Political Science,
University of Wisconsin–Madison,
author of *The Politics of Resentment*

"Jerry Apps has lived and written about farming and rural life for many years. There are few individuals who have been more steeped in the connections between—the rise and fall of—farms and rural communities of America. If Jerry thinks we've got a problem, a serious one, we should pay attention. In this concise book, he identifies what he sees as the problem and offers up 'approaches and ideas to move agriculture and its rural cities and villages forward with excitement, hope, and success.' I for one count this book as a gift."

—**Richard L. Cates, Jr.**, farmer, and
new farmer trainer, Spring Green, Wisconsin

The Speaker's Corner Books Series

SPEAKER'S CORNER

Speaker's Corner Books is a series of book-length essays on important social, political, scientific, and cultural topics. Originally created in 2005, the series is inspired by Speakers' Corner in London's Hyde Park, a bastion of free speech and expression. The series is influenced by the legacy of Michel de Montaigne, who first raised the essay to an art form. The essence of the series is to promote life-long learning, introducing the public to interesting and important topics through short essays, while highlighting the voices of contributors who have something significant and important to share with the world.

On Farms and Rural Communities

An Agricultural Ethic for the Future

Jerry Apps

Fulcrum Publishing
Wheat Ridge, Colorado

Copyright © 2024 Jerry Apps

All rights reserved. No part of this book may be reproduced, stored in a retrieval system, or transmitted in any form or by any means, electronic, mechanical, photocopying, recording, or otherwise, without the prior written permission of the publisher.

Library of Congress Cataloging-in-Publication Data
Names: Apps, Jerold W., 1934- author.
Title: On farms and rural communities : an agricultural ethic for the future / Jerry Apps.
Other titles: Speaker's corner books.
Description: Wheat Ridge, Colorado : Fulcrum Publishing, [2024] | Series: The speaker's corner book series | Includes bibliographical references.
Identifiers: LCCN 2023043189 (print) | LCCN 2023043190 (ebook) | ISBN 9781682754641 (paperback) | ISBN 9781682754658 (ebook)
Subjects: LCSH: Sociology, Rural. | Agriculture--Social aspects--United States. | United States--Rural conditions. | BISAC: SOCIAL SCIENCE / Agriculture & Food (see also POLITICAL SCIENCE / Public Policy / Agriculture & Food Policy) | POLITICAL SCIENCE / Public Policy / Agriculture & Food Policy (see also SOCIAL SCIENCE / Agriculture & Food)
Classification: LCC HT421 .A64 2024 (print) | LCC HT421 (ebook) | DDC 307.720973--dc23/eng/20231115
LC record available at https://lccn.loc.gov/2023043189
LC ebook record available at https://lccn.loc.gov/2023043190
Printed in the United States
0 9 8 7 6 5 4 3 2 1

Cover design by Kateri Kramer
Cover photo, *Wheat Harvesting in the Great North-West*, from the Biodiversity Heritage Library

Unless otherwise noted, all websites cited were current as of the initial edition of this book.

Fulcrum Publishing
3970 Youngfield Street
Wheat Ridge, Colorado 80033
(800) 992-2908 • (303) 277-1623
www.fulcrumbooks.com

On Farms and Rural Communities

I
Introduction

I am concerned about rural America and its future as the country stumbles its way into the twenty-first century, injured by the COVID-19 epidemic and unsure of itself and where it wants to head. Historically, rural communities have been the heart of this country. They continue to be essential. For the country to move forward into this century, it must have vibrant rural communities.

Agriculture has been and remains for many rural communities its core activity, and one of the most critical problems facing rural America these days is what has happened and is happening to agriculture. To put it bluntly, agriculture has lost its way. It has fallen into the trap of following the tenets of industrialization—bigger is better. Inputs and outputs. Money is more important than caring for the environment.

As I discuss the future of agriculture, I also discuss the future of rural villages and small cities. For many decades they have had a close relationship with each other. Indeed, they have depended on each other. Today, however, much of that interdependent relationship has been lost. Many rural villages and cities are struggling mightily merely to survive, to say nothing of thriving.

In the pages that follow, I will consider ways in which a new agriculture can again become the vital force it once was. Even more fundamentally, I will examine the need for all of us—rural and urban alike—to develop an appreciation for the land. How true it is that land is much more than dirt. In some depth, I will examine why we must and how we can return to a reverence for the land—no matter where we live. I will draw on the writings of Aldo Leopold, Wendell Berry, and several Native American writers who help us understand the need to look at land as a resource necessary to protect and care for—for the sake of the planet's future and for those creatures, including humans, who call it home.

Rural communities face many issues today: The direction agriculture has taken, economic development issues, political concerns, adequate rural education, rural health challenges, and more. See "Challenges Facing Rural Communities" for more information about these issues.[1]

In 2020, the vast majority of the country's population, about 83 percent (272.91 million) lived in urban areas. Some urban areas were flourishing, others less so. The remaining 17 percent (57.23 million) lived in rural areas, representing a great diversity of interests, economic opportunities, and population. Of the total land area in the United States, only about 3 percent is considered urban, while 97 percent is rural, and many rural communities are struggling mightily.[2]

I define rural communities to include the small cities, villages, farms, and open grasslands and forested lands in this country. Using the US Census definition, a rural community is everything that is not urban. So, what is the US Census definition of urban?

> For Census 2020, the Census Bureau classified as "urban" all territory, population, and housing units located within an urbanized area (UA) or an urban cluster (UC). It delineates UA and UC boundaries to encompass densely settled

> territory, which consist of core census block groups or blocks that have a population density of at least 1,000 people per square mile and surrounding census blocks that have an overall density of at least 500 people per square mile."[3]

According to the Census Bureau, the major distinction between rural and urban is the number of people living per square mile in a given area.

For practical purposes, I am including in my discussion villages and cities that have a population of five thousand people or fewer and are not closely connected to an urban area. I include all open areas as defined earlier, including farms, forests, and grasslands. The US Census Bureau estimated that about 47 percent of all the cities in the US have populations of fewer than one thousand people.[4] The state of Nebraska, for example, has about 250 villages with fewer than one thousand people.[5] Iowa has more than one hundred villages with fewer than three hundred residents.[6] In Wisconsin, 242 villages have fewer than 1,000 people, while another 213 have between 1,000 and 5,000.[7] In North Carolina, 437 villages have fewer than 5,000 people.[8]

In this book, I will share personal information about my interest in the future of agriculture, and rural villages and cities. I begin by looking at the history of rural communities, with a focus on agriculture, from settlement days to the present. I will discuss the situation of rural communities today, as they struggle to find their way with never-ending change bubbling up all around, too often overwhelming them. And finally, I will discuss possible solutions to the challenges rural communities face—especially agriculture—with some examples of success.

The first sixteen years of my life I spent on a small dairy farm in Waushara County, Wisconsin. I was born in the farmhouse where I lived those years, helping with chores and other farmwork as I got older. We had few conveniences that we take for granted today—no electricity, no indoor plumbing, and we heated the house with woodstoves. I attended a one-room country school for eight years, and then attended high school in Wild Rose (population 582). Upon graduation from high school, when I was sixteen, I received a scholarship to attend the University of Wisconsin–Madison (UW) beginning in the fall of 1951. Neither of my parents had graduated from eighth grade, and no one in my extended family had attended a four-year college (a few of them had attended the Normal School in Wautoma that prepared them to teach in one-room country schools). My attending the university was a new experience for all of us.

Arriving in Madison, and not knowing anyone, I arranged my few belongings in a single room in a rooming house on Orchard Street—I couldn't afford to stay in the dorms. Rent for my single room was five dollars a week, which remained the same for the four years I lived there. Several other students also lived in the rooming house. A student living across the hall from me offered to walk with me to a nearby drugstore that offered meals—forty-nine cents for lunch, as I recall. As we walked along University Avenue, my newfound friend, not knowing my rural background, commented, "You walk just like you're walking behind a plow." My reply to him was, "That's what I was doing a few days ago."

After that little exchange, I became concerned that my ruralness might get in the way of adjusting to university life. I decided I was going to learn how to be a city person. It began with the thought: *I must learn how to walk city.* I watched how people walked down University Avenue, and within a few days, I believed I could walk "city" with the best of them. By the time I graduated from UW in 1955, I had washed most of my ruralness out of me and I was congratulating myself on becoming a city person. At least, that is what I thought I had become.

The Korean War began in 1950 and raged during my early years at UW. Several of my fellow high school graduates were drafted into military service. I joined the Reserve Officer Training Corps (ROTC), which provided me a deferment. But upon graduating as a second lieutenant, I was required to spend time on active duty in the army. I was stationed at Fort Eustis, Virginia. There I met a group of fellow officers in a situation similar to mine. During our off time, we got to know each other well. I discovered that nearly all of them were city guys from places like Chicago, Boston, New York, and San Francisco. Although I tried to keep up my "city guy" persona, I began to wonder if that wasn't a mistake. When I heard about the growing-up circumstances of these fellows, I began to feel that my growing up on a farm in a rural community was quite special. I had learned things on the farm that suited me well in the military, not the least of which was knowing how to shoot a rifle, which I had learned how to do when I was twelve years old. I also learned how to do a job without complaining and to do it well. I learned to appreciate the food that we had to eat with nary any "I don't like that," and much more.

By the time I moved from active duty to reserve status, I had changed my mind about what I had considered the negativity of growing up "rural." Nothing wrong with growing up "city," but growing up rural had many benefits I now began to appreciate. While I was on active duty in the military, I decided that I would spend my work years trying to help rural people, farmers, and small-town residents. And that's what I have done. I began my work career as a County Extension Agent, part of the College of Agriculture, University of Wisconsin–Madison. I worked primarily with rural and small-town people. Eventually, I became a professor of agriculture in the College of Agricultural and Life Sciences at UW. Since 1994, after I took early retirement from the university, I have worked as a rural historian, writing books, giving talks, and doing radio and TV shows—all to help people understand and appreciate what farm life and rural community living was all about. In addition, I wanted to convey how important it was to the development of this country and how vital it will be to the country's future.

In 2018, I published a book titled *Simple Things: Lessons from the Family Farm.* Here are some of the "simple" things I learned from growing up on a farm. I learned the power of curiosity, to look where others did not choose to look, to listen to what others chose not to hear. I've never forgotten my father's admonition, to always look in the shadows and listen for the whispers. I learned that time on the farm is different from that tracked by clocks and watches, that it has more to do sunrises and sunsets and the changing of the seasons.

I learned that money doesn't always represent what is important—that the smell of freshly mown hay, the beauty of a sunset, and the power of a thunderstorm boiling up out of the west cannot be expressed in dollars. A number cannot describe the sheer beauty and mystery of a newborn, no matter if it's a new baby piglet, a newborn calf, or a new baby brother or sister. I learned the importance of taking care of the land, treating it as a gift that was lent to me as a farmer with the hope that I would respect it, care for it, and leave it in better shape than when it was given to me.

At my one-room country school, I not only received a solid academic education, I also learned how to stand in front of a group and "say my piece." And I learned the importance of everyone working together, sharing and caring for one another each day.

Perhaps most importantly, I learned to appreciate life's simple gifts: silence, dark nights, never giving up hope, not fearing taking risks, friends, and family.[9]

Looking back on the history of rural communities, many of them relied on mining, logging, manufacturing, or agriculture for their economic survival. Agriculture, especially food production, has been a mainstay for many rural communities from its very early days to more recent history. And for the United States to move forward into a sustainable future, it not only must have vibrant cities committed to humane living conditions for all its residents, as well as environmentally friendly approaches to economic growth, it must also have vibrant, environmentally friendly, and humane rural villages, farms, and other rural venues. The big cities of the country depend on the rural areas. The rural areas depend on the big cities. One is not more important than the other. Both are of equal importance. And, for the rural communities to do well, agriculture and the related food systems must also do well. We must all eat to survive, and thus agriculture is a foundational element.

In this book, I focus on how agriculture has changed and, in my opinion, how "industrial-size" agriculture has lost its way and is more of a detriment to the progress of rural communities than it is a help for them to succeed. I will suggest food production alternatives that fit a sustainable future for rural communities. Undergirding all of this is my concern for land use, and the importance of looking at land as more than merely an economic commodity to be exploited for maximum profits. As I earlier mentioned, I am influenced by Aldo Leopold and his ideas of following a land "ethic" when thinking about land use. I am also influenced by the Native American philosophy that making a decision today ought to be guided by the question: What will be the effects of that decision seven generations from today?

In this book, I show why a new agriculture and revitalized rural villages and cities are necessary for the US to move into the future without stumbling and falling.

II
History of Agriculture and Rural America

In 1830, 91 percent of the US population of 12,866,020 lived on farms and in small villages. America was essentially a rural country.[10] It wasn't until 1920 that more than half of the US population lived in urban areas. The country was built on its farms and villages. As recently as 1940, 43 percent of the country's population of 131,669, 275 was considered rural.[11]

In the upper Midwest, the region that I know best, early settlers depended on oxen to do the heavy work on the farm such as breaking the virgin soil, pulling crop-laden carts, and transporting people from here to there.

> The ox carried the [settlers'] burden . . . they toted them in and did the heavy work around

> the homestead after they finally got settled. They helped clear the land, did the breaking and plowing, [and] hauled the grain to market. . . . In the spring of the year, the travelers along the trails, highways and byways could hear the familiar cry echoing from every farm place, "gee," "haw.". . . The ox was a faithful critter, easily trained, docile in demeanor and tough from his horns to his hoofs.[12]

Although the oxen did the heavy work, such as plowing the fields and toting the crops from field to farmstead, much of the farm work was backbreaking, human work. These pioneer families, mother, father, and children worked from dawn to dusk, seven days a week in order to survive in this new land. It was a land of no cars, few neighbors, no trains, and few stores.

In the Midwest, many of these pioneers had come from New England—from New York, Maine, and Vermont. Many also had come from "the old country," from Germany, Poland, England, Norway, and Sweden. These early pioneer farmers were mostly self-sufficient. They grew their own vegetables and fruits, eating them fresh and preserving them for winter use. They had a few hogs, chickens, and a couple of cows for milk. They hunted game to add to their food supply.

Living conditions on the frontier in the late 1840s and 1850s were challenging and often dangerous. The winters were long, roads were nearly nonexistent, the distance to a settlement was often considerable, and the land, although often fertile, was covered with trees, which had to be removed before a crop could be planted.

> Sometimes a pioneer farmer was fifty or a hundred miles from a grist mill, a store, or a post office, and generally his highway was but a blazed bridle path through a tangled forest. Often his only entertainments throughout the year were "bees" (events where pioneers helped each other) for raising log houses or barns for newcomers and on these occasions all the settlers for scores of miles around would gather in a spirit of helpful comradery.[13]

The life of a frontier woman was especially challenging. She dispensed medicines when there was illness, served as a midwife when babies were due, and perhaps most importantly, made sure that her family was well fed. She cared for the children, was responsible for the vegetable garden, took care of a small flock of chickens, and sewed, washed, and mended the family's clothes. In the early days on the frontier, women milked the cow or two that the family owned and made the cheese and churned the butter. These frontier women often faced loneliness as neighbors were far away, and husbands were gone for extended periods of time to earn extra money. Many farm men, especially those in central Wisconsin, left for the logging camps of the north during the winter months, leaving their wives and children to survive through the long, cold winters. Many of these women had grown up in New England or Europe "in cozy neighborhoods with relatives and neighbors close at hand."[14]

In 1862, the US Congress passed the Homestead Act which provided

> that any adult citizen, or intended citizen, who had never borne arms against the U.S. government could claim 160 acres of surveyed

> government land. Claimants were required to "improve" the plot by building a dwelling and cultivating the land. After five years on the land, the original filer was entitled to the property, free and clear, except for a small registration fee. Title could also be acquired after only a 6-month residency and trivial improvements, provided the claimant paid the government $1.25 per acre.[15]

Some four million homestead claims were filed in thirty states, although only 1.6 million deeds were officially obtained. Leading states with successful claims were Montana, North Dakota, Colorado, and Nebraska.[16]

Many of the farmers in the Midwest—Indiana, Ohio, Illinois, Wisconsin, Iowa, and Minnesota—grew wheat during these years just before and for several years after the Civil War. By the 1840s, many of these wheat-growing farmers had replaced their steady but slow-moving oxen with draft horses, Belgians, Percherons, and Clydesdales; with Cyrus McCormick's invention of the reaper in 1831, a considerable amount of human labor had been replaced with horses.

Life in the country continued as it had from generation to generation with few changes. In the rural Midwest during the 1930s, the years of the Great Depression, and continuing to the end of World War II in 1945, the average size farm was 160 acres—a quarter section. Many farmers had small dairy herds, twelve to fifteen cows that they milked by hand, twice a day, 365 days a year. Their main power source was a team of horses that pulled the plows, hay wagons, and grain and corn binders. Besides the milk cows, most of these farmers also raised hogs and chickens, grew oats and corn for animal feed, and had several acres of pasture for the farm animals. They practiced crop rotation, beginning with corn, then oats, then hay crops for several years, and finally a few years of pasture before once more plowing the ground and planting it to corn.

During the 1930s and 1940s, many of the farmers also grew cash crops. For example, in Waushara County, Wisconsin, where I grew up, many of the farmers grew potatoes—as many as twenty acres as a cash crop. Most families also had a small cucumber patch, up to an acre depending on the size of the family, as it was the farm kids that did most of the cucumber picking. By the late 1940s, several farmers also grew snap beans as a cash crop, at the time all picked by hand.

Farms were located about every half mile. For example, in my case, Bill Miller's farm was a half mile south of our farm. Griff Davies was a half mile east, Alan Davis a half mile north, and Andrew Nelson and Charlie George, living across the road from each other, a half mile to the west.

Each farm was mostly self-sufficient, growing the feed necessary for the farm animals and producing the fruits, vegetables, and meat that the family preserved for its own use. Although each farmer cut his own grain, when the threshing machine arrived in the community, all the farmers worked together as a threshing crew moved from farm to farm. The threshing crew hauled the oat bundle shocks from the farmer's fields, and fed them into the machine that separated the oat kernels from the straw. Neighbors also helped each other with silo filling, corn shredding, and wood sawing. By the 1930s, most of the farmers had telephones. If someone was hurt, if there was a fire, or a windstorm raised havoc, the party-line telephone was used to call for help. And help came as quickly as possible as these self-sufficient farmers were highly dependent on each other. The idea of community and neighbors grew out of these relationships. Everyone in the immediate neighborhood knew each other, including knowing the children's names, and even the names of each other's horses and farm dogs.

The women helped each other during the various community bees (when farmers helped each other) by assisting with the preparation of the meals—which were akin to meals served during celebrations such as Thanksgiving and Christmas. The farm women, especially those living in the upper Midwest where the winters were long and fierce, suffered from being alone for long stretches of time. The men, by hauling their milk to a nearby cheese factory, or taking their grain to be ground to the village grist mill, had an opportunity to chat with other men as they waited their turn to unload their milk or for their grain to be ground. To overcome their loneliness, the neighborhood women often gathered to make quilts, a project that required them to meet several times to complete the task. It offered an opportunity for them to visit with each other on cold, lonely winter days.

Land Use and Abuse

Because of the Homestead Act and a general increase in the country's rural population as people moved from New England west, and as new immigrants arrived, thousands of acres of land came under the plow and were planted to wheat. Wheat prices were high during the mid-1800s, encouraging even more grassland to be plowed.

And then wheat prices fell and the once-thought dependable rains quit falling, and widespread drought rolled over the middle of the countryside starting in the 1930s. Thousands of acres of homesteaded land in the southwestern states—the Oklahoma Panhandle, southeastern Colorado, the western third of Kansas, as well as the Texas Panhandle and northeastern New Mexico—suffered the most when an extended drought overtook the region during this decade. With no grassland roots to hold the soil, dry winds sent the fragile topsoil high into the air, causing blinding dust storms and destroying farms and farmsteads. The phenomenon was known as the Dust Bowl, and it occurred at the same time as the Great Depression. Farms were abandoned. Farmers, along with thousands of others, moved on, trying to find work, trying to find something to eat. By 1934, an estimated thirty-five million acres of once rich farmland was considered useless for farming. Another 125 million acres were rapidly losing their topsoil.[17]

The drought also found its way into the upper Midwest. As a little kid, I remember the clouds of dust boiling up out of the west, sifting into our farmhouse under the windowsills—a fine, dirty dust that represented the topsoil of the farms in central Wisconsin, moving east, devastating once rich farmland.

President Franklin Delano Roosevelt sponsored several governmental programs designed to help the poor and displaced farmers as well as other unemployed people. In 1935, Congress established the Soil Erosion Service, which became the Soil Conservation Service and is now known as the Natural Resources Conservation Service. President Roosevelt and Congress also passed legislation in 1933 creating the Civilian Conservation Corps, designed for out-of-work young men who worked on various conservation projects. They planted windbreaks and encouraged contour farming to help prevent soil erosion. These young men planted thousands of trees where loggers had clear-cut the forests and left the land to erode. They also helped create and improve national parks and forests.

Following the Great Depression came World War II, which began on December 7, 1941, and continued until 1945. Farm prices increased a bit, but rationing of such things as tires and gasoline put further restraints on the nation's farmers. People were exhausted after a decade of depression and four years of war. Starting in 1946 great changes—some might say revolutionary changes—began occurring in rural America.

III
A Personal Example of Early Farm Life

Before the late 1940s, with two exceptions, no one in my home farming community (Township of Rose, Waushara County, Wisconsin) had electricity. The population of the township, which was six miles square, was 529 people at that time. The exceptions to no electricity: Andrew Nelson had a Delco Wind charger, a wind-powered generator that charged a small roomful of batteries and provided some 32-volt electricity—mostly light bulbs. Bill Witt, my grandfather, had a Delco gasoline-powered generator, with several batteries that provided 32-volt electricity.

For water, everyone had a well, and for most of the farmers—our farm included—a windmill-powered pump. Some of the farmers were able to afford gasoline engines that powered their pumps. No one in the community had indoor plumbing, and everyone cooked their meals and heated their homes with woodstoves. A few of the farmers had battery-powered radios—we did—and almost everyone had a party-line telephone.

One of the clear differences between those living on farms, and those living in villages such as Wild Rose, were the conveniences that we all take for granted today. For example, by the 1930s, people living in Wild Rose had electricity, provided by the grist mill, which had a water-powered generator.

All the children in our neighborhood attended the Chain O' Lake, one-room country school. Everyone walked to school, some of the students as far as two miles. For me, the walk was only about a mile.

The school, when I began attending in 1939 as a five-year-old first grader, had no electricity, no indoor plumbing, and was heated with a woodstove—no different from nearly all the farm homes in the community. The school had a battery-powered radio, which allowed us to connect to the Wisconsin School of the Air with radio programs from radio station WHA, specifically designed for rural schoolchildren. Examples included *Afield with Ranger Mac* (a nature program with Wakelin McNeil), *Let's Sing* with Professor Gordon, and *Let's Draw* with Jim Schwalbach.

In 1946, my father, working with Henry Haferbecker, the Waushara County agricultural agent, organized a 4-H club in the Chain O' Lake School District. In the early years of the club, it had about ten members, all students at the one-room school. Four-H provided a wonderful opportunity for farm kids, both boys and girls, to learn topics that were outside of what they were learning in school. For example, in my first year of 4-H, I enrolled in the dairy calf project and the forestry project. I learned the importance of keeping careful records of what I was feeding my 4-H calf as well as a record of how I was tending my little pine tree nursery. I learned how to teach my calf to lead—a necessity as I had to lead him in the show ring at the county fair. In the process I learned the importance of patience and perseverance. I learned *Robert's Rules of Order*—a book that shows how to run a meeting in a democratic way, including how to make motions and vote on them. These are lessons that I have used throughout my life. I learned what fun it was to spend several days at the county fair in Wautoma, staying overnight in a tent and caring for my 4-H calf during the day.

I spent the first sixteen years of my life on the home farm, learning the ways of farm life and rural living. Absorbing the importance of work and how things had to be done right, on time, and with no complaining, at least not while your dad was in hearing range. I learned how to make do with little during the years of the Great Depression. I learned how not to be jealous of my city cousins who had electricity, indoor plumbing, and central heating in their homes years before we did. We had none of these until 1947 when electricity arrived. Indoor plumbing and central heating arrived some years after I left home.

One of the most important things I learned from those sixteen years of growing up on a farm was something we never talked about but knew was of vital importance: the land—the importance of the land and how it needed to be respected and cared for. How our livelihood depended on it. Our farm, in the vernacular of the farmer, did not have good land. Meaning that in other parts of the state and country there was better land, more fertile land, less hilly land, less stony land. But we had what we had, and by the late 1940s, my dad owned our farm, free and clear. No mortgage, no interest payments. Only annual taxes. It had always been his goal to be debt free, and he was. And although he never talked about it, I knew he was proud of what he had accomplished. He had been in a partnership with the land since 1924, when he and my mother first rented, and then bought, the home place, farming there until 1973, when he sold it.

The relationship my dad had with the land was more than economic, although that was an important part of it. He would never have used these words, but I believe he had a spiritual relationship with the land. I remember so well, often on a summer Sunday afternoon, he would hike around the farm, and sometimes my twin brothers and I would hike with him. He said he was checking on the fences and the crops and the grazing livestock, and he was. But there was more. Sometimes he would just stand looking and not say anything. And then we would walk some more, perhaps to a hilltop where he would stand once more and gaze at the vista in front of him, not saying anything. And he wanted me to do the same: not talk, just look. And I did. I believe my dad was a firm believer in what Aldo Leopold called a land ethic—the relationship of people to the land. See part VII for more about Leopold's writing on this topic.

IV
Rural Villages in an Earlier Day

Just as the neighbors in a farming community depended on each other, farmers also depended on the nearby village. And the village depended on the farmers that surrounded it. The nearest village to our farm was Wild Rose, about four and half miles to the east. Wild Rose was located on the Pine River where a millpond and mill had been built in the 1870s.

During the 1930s, the following business places could be found in Wild Rose: cheese factory, grist mill, sawmill, mercantile (a general store plus opera house and dentist office), meat market, hardware store, drug store, blacksmith shop, harness maker, barber shop, furniture store and undertaker, two taverns, hotel, restaurant, lumberyard, auto repair garage, several potato warehouses, three gasoline stations, post office, and a bank. The village also had an elementary and high school, plus a Methodist, Presbyterian, and Baptist church. By the late 1890s, Rural Free Delivery meant mail was delivered to each farmer's mailbox. Prior to that time, the post office was in one of the general stores, which is where farmers got their mail.

Most of the farmers in the Wild Rose area sold their milk to the cheese factory, had their cattle feed ground at the grist mill, had their horse harnesses repaired by the harness maker, bought their lumber at the lumberyard, their hardware (nuts, bolts, etc.) at the hardware store, and their clothing at the mercantile or other general stores. Farm wives bought their basic cooking needs such as flour, sugar, coffee, and salt at one of the general stores.

Clearly, there was an economic relationship between the rural villages and the farming communities surrounding them. But the nearby village was also a cultural center for the area. Farmers had essentially three sources for cultural opportunities: their churches, the one-room schools, and what the nearby villages offered.

The village of Wild Rose offered several types of cultural opportunities. The village traces back to 1873, when it was first organized. In 1901, the Chicago and Northwestern Railroad built a rail line from Fond du Lac to Marshfield, with a stop at Wild Rose. Shortly after that, Tom Patterson erected a two-story brick mercantile building that was multipurpose. In its basement were bowling alleys; on the first floor was a general store with dry goods, clothing, and groceries. The second floor included offices and an opera house. At that time, the late 1800s and early 1900s, opera houses offered a wide variety of programs, including lectures, musical events, and plays. Opera houses had become popular, especially in the Midwest. Although these venues were called opera houses, no high opera was performed. The majority of the musical productions were based on the popular music of the day.

The Chautauqua, tracing back its beginnings to New York State in 1874, expanded its reach to rural areas in the country. In 1915, when the Chautauqua movement was at its peak, some twelve thousand rural communities hosted Chautauqua programs. After Teddy Roosevelt saw one of the Chautauqua performances, he said it was "a source of positive strength and refreshment of mind and body to come to meet a typical American gathering like this—a gathering that is . . . typical of America at its best."[18]

I remember as a kid attending one of the Chautauqua stage presentations that had come to Wild Rose. It was the first time that I had ever seen professional actors perform on a stage—my only experience was with the little plays that we performed at our one-room country school. I don't remember the name of the play, but it had a western "shoot-'em-up" theme with real pistols and gunfire that scared the bejabbers out of me. Later, Pa told me the bullets were not real.

Starting in the 1930s and continuing into the 1940s and 1950s, many rural villages offered free movies during the summer months. My family never missed a free outdoor show, offered every Tuesday night. The businesspeople in Wild Rose sponsored the shows. The outdoor theater consisted of rough sawed planks nailed to blocks of wood and lined up on the banks of the millpond. A huge willow tree provided a place for the "movie man" to fasten a big white bedsheet that served as a screen. The movies were usually Westerns featuring such stars as Gene Autrey, Roy Rogers, and assorted "sidekicks."

When the milking was done, my mother, dad, two brothers, and I would pile into our old 1936 Plymouth and head toward Wild Rose for this special night. If my brothers and I were lucky, Pa would give each of us a dime that we could spend on a double-dip ice cream cone, a real treat as we had no electricity at home and thus no place to keep ice cream.

We would usually arrive in town just before dark. The movie man always waited for dark before he turned on the projector—a huge piece of equipment that filled an enclosed trailer. We knew the movie was about to start when Specks Murty, the village marshal, screwed loose the streetlight in back of the outdoor theater, casting the entire area into darkness.

As we watched the film flickering on the bedsheet screen, we could listen to the bullfrogs calling from the millpond, and occasionally we might hear a whippoorwill calling its name over and over again from the Roberts farm located on the other side of the millpond. When the show was over, we once more climbed into the old Plymouth and headed home, a bit sleepy because "show night" meant we probably didn't get to bed until after ten, and we knew we had to be up in the morning at least by five thirty.

The free movies, along with the other cultural offerings, such as the Chautauqua and the dances and talks offered at the opera house, added an important dimension to the lives of the farmers who lived in the community. For many years the economic and cultural interaction between the farmers and the village people would continue—many believing that the relationship would continue indefinitely. But it was not to be.

V
Dramatic Changes in Rural America

The Depression years of the 1930s hit the country's rural communities hard, as it did the urban areas.

Prices for farm products fell to levels not ever seen. Farmers, different from those living in the big cities, had food to eat and work to do. But farmers had other challenges. I interviewed my father when he was in his eighties. This is how he remembered the Depression years on our farm:

> In 1930, I had eight milk cows and maybe four or five head of young stock. Four horses were my source of farm power. We sold cream and butter in those days. We got fourteen cents a pound for cream. My cream checks were three dollars and fifty cents a week. We always had a couple hundred chickens—eggs sold for seven cents a dozen. Consumer goods were cheaper to buy, too. A pair of shoes sold for one dollar and fifty cents. Bib overalls—seventy-five cents, a blue work shirt—twenty-nine cents, and often on sale for twenty-five cents. I paid seventy-six cents for a pair of leather gloves, and a three-tine pitchfork cost seventy-five cents. I bought a steel-wheeled wagon for seventy-five dollars, which was a lot of money. Ma and I moved to the farm in 1924, renting from John R. Jones.

We saved every dollar we could with hope of one day buying the place. During the Depression the bank closed, and we only got thirty-five percent of the money we'd saved. We did finally buy the farm in 1936.

During the worst of the Depression we always had a few pigs, mostly Chester Whites and Durocs. Most of the time we butchered them ourselves and then peddled the meat. We sold the back half for ten cents a pound and the front half for eight cents a pound. We got forty cents for a pig head. We mostly peddled in Wisconsin Rapids because the people working in the papermills still had some money. When the war started in 1941, pork went from six cents a pound live weight to 10 cents a pound in one year. One time I had a hundred pigs, real mortgage lifters.

During the 1940s, the war years, I began selling our milk in cans. A milkman came by to pick it up every day, and we got around seventy cents to eighty cents a hundred pounds for it.

In terms of living conditions, a considerable cultural gap existed between farmers and those living in the villages and cities. That cultural gap was defined with one word—"electricity." By the 1930s, 90 percent of urban dwellers had electricity, while only 10 percent of rural dwellers did. Farm organizations such as the Farm Bureau and Farmer's Union had been advocating for electricity on the farm for several years, and in 1935, during the depths of the Great Depression, President Roosevelt signed an executive order establishing the Rural Electrification Administration (REA), which provided loans to local electric cooperatives. Almost immediately, these co-ops began organizing all over the country, often with the help of county agricultural agents.[19]

The REA was poised to bring electricity to rural locations. But after the United States entered World War II in 1941, the newly organized REA cooperatives were unable to obtain necessary supplies, such as copper wire or sufficient labor. It was not until after the war that many farmers would see electricity, with all its benefits, come to their farms.

With electricity on the farm, a huge cultural gap between rural and urban dwellers closed. Electrical suppliers quickly learned that their farm customers used far more electricity than the suppliers had anticipated. For farmers, the light bulb was probably the least of what they prized. It was electric motors, electric heaters, milking machines, silo unloaders, milk refrigeration units, and more that farmers found so convenient and labor saving.

At a deeper level, farmers' lives had long been seen—at least by some of their urban neighbors and relatives—as less rewarding and fulfilling because they had no electrical power. Now farmers had access to the same electrical conveniences as people living in the cities.

According to the report *Wisconsin Agriculture in Mid-Century*, prepared in 1953 by the Wisconsin State Department of Agriculture, electricity was "one of the best and cheapest workers on Wisconsin farms. . . . It is used in the performance of numerous farm tasks as well as providing light, power and fuel for the farm home. The water that flows in the barn is often supplied by an electric pump. Electricity is used for the milking machines, milk coolers, feed grinders, fans, radios, washing machines and many other items."[20] By 1950, 93 percent of the farms in Wisconsin had electricity, paying an average monthly bill of $9.74.

After replacing oxen in the mid-1800s, horses continued to handle the heavy work on the farm through the Depression years. Many farmers didn't purchase their first tractor until after World War II. During the Depression, when gas tractors became available, many farmers simply couldn't afford them. (A handful of farmers owned steam tractors, mostly used for powering threshing machines.) Most farmers were accustomed to farming with horses and enjoyed working with the animals.

In the 1950s, better farm prices allowed more farmers to buy tractors. With tractors came a new generation of tractor-powered farm equipment. Dairy farmers needed a barn full of hay to feed their hungry cows throughout the long winters, and hay making with horses was slow, tedious, dusty, and dirty work. Thankfully, tractor-pulled hay balers allowed the farmer to forgo bunching newly raked hay by hand, pitching it onto a horse-drawn wagon, and then forking it into the haymow.

With tractors and newly developed forage harvesters, corn binders were eliminated as well as silo fillers. With a tractor and a mechanical corn picker, corn shredding bees (when farmers helped each other) were no more. Tractor-pulled grain combines eliminated the need for grain binders for shocking grain by hand, and for the threshing crews that moved from farm to farm. In the wake of these technical innovations, communal work gatherings such as silo-filling bees, corn shredding bees, and threshing bees disappeared. These work bees not only helped neighbors with the harvest but were social events and a way to tie rural communities together.

These several changes in farming practices had profound structural effects on rural communities. The most obvious was the decline in the number of farms. With tractors, electricity, and new disease-resistant and higher-yielding crop varieties, Wisconsin farmers could grow more crops and milk more cows with less human power than ever before. As a result, farmers—and especially farmers' children—had fewer opportunities in farming. Starting in the late 1940s, with more young people having completed a high school education, many of them left the farms for work in the cities. Some of the more fortunate young people left to attend college. The number of farms in the United States peaked at 6.8 million in 1935. The number plummeted to 2.8 million farms by 1973 and continued to fall, reaching 2.02 million in 2020.[21]

As the number of farms shrank, farms themselves became larger, the farm population declined, and rural communities changed dramatically. One-room country schools, a prominent feature in all Wisconsin communities, mostly closed by the mid-1950s, as school buses transported the remaining young people on farms to the consolidated schools in the nearby villages and cities. Many country churches closed as well. Crossroads cheese factories ceased operations as trucks and improved roads made travel from farm to market easier. And with better roads and better cars, people traveled to large retail centers for their purchases, causing smaller villages to lose business.[22]

Contributions of Rural America to Urban Centers

Often not discussed are the contributions that rural America made to the cities following World War II and for several decades after. As small family farms disappeared, the young people that were born and raised there—I am one of them—were forced to find work in an urban center. They took with them a set of beliefs and values that generally served them well in nonfarm settings. Farm kids knew how to work, knew how to follow directions, knew how to do what needed doing, and tried to do so to the best of their abilities. There were exceptions—but not many.

I remember one time when I applied for a summer job working at a pea cannery. My interview with the hiring person was brief. His first question, "Are you a farm kid?"

"Yes," I answered.

"Do you have a crescent wrench?"

"No, but I can borrow one from my dad."

"You're hired."

There were no further questions. The fellow obviously had had experience with farm kids. The job was tough, hard physical work, sometimes with twelve-hour days. But in two months, it was over. I learned a lot from the experience.

VI
Rural Communities Today

First, let's look at how land is used in the United States today. Note that more than half of the country's land (51 percent) is devoted to forests and what is called shrubland, which includes desert lands. Agriculture, grasslands, and pasture lands make up 34 percent.

2020 Land Use in the United States

- Forests—27 percent
- Shrubland—24 percent (includes desert lands)
- Agriculture—17 percent
- Grasslands and Pasture—17 percent
- Wetlands—5 percent
- Other—5 percent
- Open Space—3 percent
- Urban Areas—2 percent

Note that 98 percent of the land mass in the United States is considered rural, with but 2 percent devoted to the country's cities. However, as noted earlier, in 2018, urbanities made up more than 82 percent of the country's population. More than a quarter of the country's land mass is in forest land, divided about equally between deciduous and evergreen trees. Seventeen percent of the land is devoted to agriculture along with another 17 percent in grassland, some of it used for agriculture.[23]

Rural America Today

Wheat, corn, and soybeans are the major crops grown in the United States. Cotton is also a profitable non-food crop. Soybeans, corn, and wheat are exported to countries such as China and Japan. Corn is a crop with multiple uses, from food to ethanol (added to gasoline as a fuel). About 40 percent of the US corn crop is used for livestock feed.[24]

In 2021, there were 2.01 million farms in the United States, down from a high of 6.8 million in 1935, the peak number for the US. In 2022, 895 million acres of land were in farms, with the average farm size 445 acres. The dramatic drop in farm numbers was driven by improved animal and crop genetics, farm technology, and increased efficiencies. Even with the decline in farm numbers and farm population, total farm output has continued to increase. Corn and soybeans accounted for more than 40 percent of all US crop income in 2020. In animal agriculture, Cattle and calf receipts accounted for 38.2 percent of the income from animal products. Dairy products accounted for 24.6 percent, and poultry and eggs contributed 21.5 percent. Although 89 percent of farms were considered small (less than $350,000 gross annual income), large-scale farms (earning more than one million dollars per year) accounted for about 3 percent of the farm numbers, but produced 46 percent of total farm income.[25]

The Importance of Rural Communities

As they have for many years, rural areas continue to provide all Americans with water, food, energy, and recreational opportunities. Historically, rural community economies have relied on agricultural, mining, manufacturing, and service industries such as recreation (fishing, hiking, bird-watching, kayaking, etc.). These vary from rural community to rural community, with reliance on agriculture being the most important for many.

The Center for American Progress researchers concluded,

> Though some are thriving, rural areas overall have yet to match the employment level reached prior to the 2008 recession, and deep poverty persists in many rural communities. Beyond barriers to jobs and economic opportunity, some rural areas also lack access to crucial services such as health care and broadband internet. Rural America is not homogenous and should not be discussed or treated as such. In order to properly address the issues facing rural communities across the country, advocates and policymakers must understand the diverse nature of rural communities and the various systemic challenges they face.[26]

Many residents of rural communities live in the country and commute to nearby cities for employment. Andy Lewis, who spent many years working as a University of Wisconsin Extension Agent in rural communities, remembered this: "My administrative assistant's husband was a carpenter who commuted from Fennimore (population about twenty-five hundred) to Madison for the higher wages. However, that ninety-minute one-way commute meant he was spending fifteen hours a week, sixty hours a month driving. That meant he had sixty hours less time for volunteering or for family. Many farm families took second jobs for the health insurance, which was critical for the dangerous occupation of farming."[27]

Characteristics of Rural Communities in 2021

- A small population size
- A generally low population density
- A lower cost of living
- Lower wages and, consequently, higher poverty rates
- Considerable farmland, ranchland, and nature areas
- Less access to shopping, doctors, and other services
- An aging population[28]

A myth associated with rural America is that it is all white and entirely agricultural. The truth: rural America is increasingly diverse.

One in five Americans lives in a rural community and more than one in five (22 percent) of rural residents are people of color. Rural Native American, Asian, and Latinx groups are growing fastest, followed by African Americans, with modest population gains, and non-Hispanic white groups experiencing the slowest growth. Most rural Americans are not farmers—in fact only about 6 percent of rural Americans are employed in agriculture. The largest employers are education, health care, and social assistance sectors followed by retail, construction, and transportation. Rural communities exist in nearly every state and territory, not just the Midwest.[29]

Advantages and Disadvantages of Rural Living

Advantages of rural living include more open spaces; less crowding; generally quieter surroundings with few sounds of trucks, sirens, and automobiles; less crime; and a slower pace of living with lower levels of anxiety. Disadvantages include poor employment opportunities, lack of basic services, inadequate to no public transportation making it difficult for people without cars to get to work, lack of high-speed internet, and lack of sufficient numbers of hospitals, medical clinics, and health-care professionals. A related challenge to many rural communities is the increasing number of older adults with associated health-care problems. Besides a growing older population, many rural communities are also experiencing a general decline in number of school-age children, resulting in school closings and the need to transport children greater distances to schools.

Poverty is a problem in many rural communities. One of the reasons for rural poverty is young people leaving the area for jobs in an urban area upon graduating from high school. Compounding the poverty problems, rural areas often lack human service programs designed to assist the poor.[30]

Population Trends in Urban, Suburban, and Rural Communities

During the 2010s, rural areas tended to slightly lose population (-0.6 percent) while urban areas gained

(8.8 percent). Rural counties that were not considered poor increased their population slightly (0.1 percent). Those that were persistently poor declined by 5.7 percent. This was not true for poor urban counties. Urban counties both poor (5.8 percent) and not poor (8.9 percent) gained population. The nature of the rural economy also affected population trends. Rural states with the highest percentage of a declining population "were relatively more dependent on farming . . . manufacturing . . . and extraction of coal, gas, oil, and other natural resources."[31]

An Example

In looking at those rural communities that had a close interrelationship with their nearby villages, here is what happened to one—my hometown of Wild Rose, Wisconsin—when agriculture went through dramatic changes following World War II. The village was incorporated as a village in 1873 and served its surrounding rural community for more than a hundred years. In 1950, the village, with a population of about six hundred, offered these businesses and services: a gristmill, a cheese factory, a feed store and lumberyard, the Midland Cooperative (which sold grease, oil, fertilizer, and farm seed), a small hotel, three taverns, several churches, a high school and a consolidated grade school, a hospital, a welding shop, a hardware store, a sawmill, a cucumber salting station, two small restaurants, a mercantile that sold everything from shoes to groceries, an ice-cream shop, a pharmacy, a dentist's office, a harness shop, a clothing store, a butcher shop, a bank, a post office, two car dealers that also did auto repairs, and a small public library.

Many farmers quit farming in the Wild Rose service area in the years following World War II. Some retired. Some sold their farms and moved to an urban area with better employment opportunities. My parents, Herman and Eleanor Apps, sold the home farm in 1973 and moved into Wild Rose when Dad retired. After the sale, the home farm was no longer farmed and would eventually be sold to seven different families, several of whom built homes on the property. With one or two exceptions, all of the neighboring farms met a similar fate. As farmers left the land, the need for village services diminished. The village of Wild Rose changed as the countryside changed.

Several farmers left the land and sold their farms to people who wanted to live in the country but who now commuted to work in distant cities: Stevens Point, Appleton, Neenah, Menasha, and Oshkosh. With the farmers gone, the village of Wild Rose changed, and changed dramatically. The gristmill closed and became a home. The cheese factory closed and was torn down. The mercantile became an antique store. The car dealers closed, as did the harness shop, dentist's office, clothing store, butcher shop, feed store/lumberyard, ice-cream shop, and pharmacy. The Midland Cooperative became a convenience store that sold gas, a few groceries, and other necessities. A second gas station remained open. The taverns continued as did three restaurants and a small grocery store. The high school and grade school continued. Two nursing homes were added. The hospital remained open, making it the only one in Waushara County. Patterson Memorial Library expanded, added a community room and kitchen, and served as the community's meeting center.

Better than many small farm-service villages, Wild Rose—because of its proximity to many lakes with surrounding cottages and summer visitors—moved from being primarily a village serving farmers to one relying on tourists. But the transition has been a struggle. The population of the village has grown to more than seven hundred, but the village is challenged to provide basic services with a limited tax base. Also, with few employment opportunities, many young people leave the community upon graduating from high school. Nonetheless, with strong local leadership and vibrant service clubs, the village is facing the future with enthusiasm, as seen in the several communitywide events held during the year that attract hundreds of people to the village.

Many rural villages and small city Main Street businesses suffered with the coming of big box stores such as Walmart and chain stores such as Family Dollar. This trend began in the 1960s and continues to this day. In the year 2,000, Walmart had 3,989 stores scattered across the country.[32] In 2023, there were 8,384 Family Dollar stores in the United States, many located in small, rural villages.[33]

Not too different from several of my neighbors, my wife and I bought one of the abandoned farms west of Wild Rose in 1966, where we spent weekends and vacation time with our three children, and now grandchildren. We operate the 120-acre place as a tree farm. We also grow a big vegetable garden there, and currently are restoring an eight-acre prairie to what it was in 1867, when it was a big bluestem grass prairie. We also practice sustainable forestry, working with professional foresters.

VII
Land Is More Than Dirt

Land is not merely a patch of ground on which to build a building, grow a crop of corn, dig for coal, or drill for oil. Land is much more. On the one hand, non-Native Americans have viewed land almost completely based on economics: What is the best use for a piece of land to bring in the most money? On the other hand, Native Americans view life as a series of interconnections; every decision has physical, economic, social, and spiritual consequences. When making a decision, each of these impacts is carefully considered. Ivan Makil, of the Pima-Maricopa Indian Community, states, "We are taught to think about how a decision we make about our land will affect the next seven generations," he says. "It is all about sustainability—about making decisions that ensure that our land, air, and water can support all forms of life for generations to come. While each American Indian tribe is unique, all tribal people believe in balancing the economic impact of a decision with its physical, economic, social, and spiritual implications."[34]

The Seventh Generation Principle is based on a Haudenosaunee (Iroquois) philosophy dated between AD 1142 and 1500. The essence of this philosophy is that the decisions we make today, especially those related to energy, land, water, and natural resources, should result in a sustainable world seven generations into the future. The principle can also be applied to relationships (between individuals, communities, and organizations)—they should all result in sustainable relationships for seven generations. Seven generations equal about 140 years.[35]

Another Native American view considers all of nature as family. "Native American teachings describe the relations all around—animals, fish, trees, and rocks—as our brothers, sisters, uncles, and grandpas. . . . These relations are honored in ceremony, song, story, and life that keep relations close," writes Winona LaDuke, Anishinaabe writer, economist from the White Earth reservation in Minnesota, and executive director of Honor the Earth, a national Native advocacy and environmental organization.[36]

Thinking of land as a family member and considering the phrase that is often used—Mother Earth—gives us another perspective on land use. Robin Wall Kimmerer, an enrolled member of the Citizen Potawatomi Nation, writes in her book, *Braiding Sweetgrass*, "It was for me, the one thing that was not forgotten, that which could not be taken by history: the knowing that we belonged to the land."[37]

Aldo Leopold, writing in the late 1940s, said this:

> Land . . . is not merely soil; it is a fountain of energy flowing through a circuit of soils, plants, and animals. Food chains are the living channels which conduct energy upward; death and decay return it to the soil. The circuit is not closed; some energy is dissipated in decay, some is added by absorption from the air, some is stored in soils, peats, and long-lived forests; but it is a sustained circuit, like a slowly augmented revolving fund of life.[38]

Aldo Leopold dug deeper into his discussion of land. He talked about the need for a land ethic, writing about the sequence of ethical development. (Ethics deals with morality, and an inherent sense of what's right and wrong.) The first level of ethics deals with the relationship between individuals. The second level of ethics is broadened to include the relationship of individuals to society. Leopold then goes on to argue for a third level of ethics—the relationship of people to the land.

> The Golden Rule tries to integrate the individual to society; democracy to integrate social organization to the individual. There is as yet no ethic dealing with man's relation to land and to the animals and plants which grow upon it. . . . The land-relation is still strictly economic, entailing privileges but not obligations. The extension of ethics to this third element in human environment is, if I read the evidence correctly, an evolutionary possibility and an ecological necessity. . . . The land ethic simply enlarges the boundary of community to include soil, water, plants, and animals, or collectively,

> the land. . . . In short, a land ethic changes the role of *Homo sapiens* from conqueror of the land-community to plain member and citizen of it. It implies respect for his fellow-members, and also respect for the community as such.[39]

But you say, "The land is something only farmers need to be concerned about—I live in the city. Why should I care about the land beyond it providing a sturdy foundation for my house, condo, or apartment building?" Reading Leopold carefully, along with Native American writing, one cannot help but conclude that just as we are responsible for acting morally, as individuals, toward each other, and acting morally toward a community of individuals, we are obligated to act morally toward the land. Every one of us has this responsibility. We may carry out our obligation in different ways, but we nonetheless have the responsibility for doing it.

I would not be the first to say that this ethical understanding of our relationship to the land to include, as Leopold says, soil, water, plants, and animals is too often missing and if present, is too often overshadowed by economic (potential profit) considerations.

VIII
Rural America and Climate Change

Climate change is a major threat to rural America, especially to agriculture, forest resources, and ultimately the rural economy. Climate change is defined as

> a long-term change in the average weather patterns that have come to define Earth's local, regional and global climates. These changes have a broad range of observed effects that are synonymous with the term. Changes observed in Earth's climate since the early 20th century are primarily driven by human activities, particularly fossil fuel burning, which increases heat-trapping greenhouse gas levels in the Earth's atmosphere, raising Earth's average surface temperature. These human-produced temperature increases are commonly referred to as global warming. Natural processes can also contribute to climate change, including internal variability (e.g., cyclical ocean patterns

> like El Niño, La Niña and the Pacific Decadal Oscillation) and external forcings (e.g., volcanic activity, changes in the Sun's energy output, variations in Earth's orbit).[40]

The results of climate change are rising sea levels and extreme weather, including hurricanes, devastating tornados, intense rainstorms, prolonged heatwaves, droughts, floods, more frequent and intense wildfires, insect outbreaks, and crop failures.

The warmest years on record in the US were those from 2005 to 2020. In 2020, the Southeast was extremely wet, and the Southwest experienced a drought. Wildfires burned millions of acres in the Northwest. "Across agricultural sectors and rural America, climate impacts contribute to an increase in invasive species and additional costs for weed and pest control, prevented or reduced plantings, decreased health in crops and livestock, production and associated income losses and damages to buildings, equipment and land."[41]

Ironically, agriculture is not only adversely affected by climate change, it is also a contributor to it. Of all the economic sector contributors to global greenhouse gas emissions, agriculture, forestry, and related land uses contribute 24 percent. Greenhouse gas emissions from agriculture come mostly from crop cultivation, livestock production, and deforestation. Other contributors to global greenhouse gas emissions include electricity and heat production, (25 percent), industry (21 percent), transportation (14 percent), other energy (10 percent) and buildings (6 percent).[42]

The United Nations Foundation offered this:

> The reliable access to safe, affordable, and nutritious food—is inextricably linked to predictable climate and healthy ecosystems. Climate change and associated severe weather, droughts, fires, pests, and diseases are already threatening the production of food around the world. Unless we act decisively, these problems will worsen, the poorest and most vulnerable will suffer disproportionately, and instability will increase. . . . In addition to being affected by the impacts of climate change, agriculture is a major contributor—and a potential solution—to climate change. In a recent report, the Intergovernmental Panel on Climate Change found that more than a third

> of global greenhouse gas emissions come from the production, distribution, and consumption of food. When it comes to producing food, the majority of agricultural emissions are related to raising livestock, followed by rice cultivation and the production of synthetic fertilizers. Moreover, as forests and grasslands are converted for agriculture, the world is losing vitally important ecosystems that remove greenhouse gases from the atmosphere. To avoid the most devastating impacts of climate change and meet the goals of the Paris Agreement, we must radically transform our agricultural systems.[43]

Unfortunately, climate change has become a contentious issue. Even with an ever-increasing body of well-researched knowledge and overwhelming agreement by the scientific community about the effects of climate change, many deniers loudly proclaim that global warming and climate change do not exist.

Policymakers, researchers, and educators face a difficult challenge as they work to inform the US citizenry about climate change's impact on the nation's agriculture (as well as cultural and economic effects beyond agriculture). Minimally, new crops and crop varieties and agriculture management strategies will be needed as the climate continues to change.

Sierra magazine editor, Jason Mark, summed up the climate change situation this way: "Rising temperatures aren't leading to reliable shifts, like winter slipping into spring, and spring folding into summer. The global climate is a finely tuned system, and even small changes in temperature—as we're witnessing—have unexpected consequences. Climate change is above all, chaotic, unpredictable and disorienting."[44]

For the future of the planet, climate change cannot be ignored. It must be of major concern for everyone, urban and rural.

IX
A Broken Food System

America, from its earliest settlement days, has depended on its rural acres and their farmers to feed the nation. As we think of the future of rural America, we must also consider the future of agriculture because rural

America and agriculture have long been linked together. Providing a ready supply of safe, healthy food has long been the goal of rural America and its many farms. But because of the mantra "bigger is better," and because of national and often local policies that encourage industrial-size agriculture, the number of farms has decreased dramatically since 1935 when the number was at its highest. As noted earlier, in 1935 the US had 6.8 million farms; in 2021 it had just over 2 million farms. With farm numbers declining, existing farms became larger; the average size farm in 1935 was 155 acres, in 2020 it was 444.[45]

Although other job opportunities exist in rural communities, agriculture and food production are uniquely rural. Agricultural rural areas are largely found in the heartland but also include much of the remainder of the country. One of the major challenges facing agriculture is the concentration of production, processing, and distribution of agricultural products in but a few huge corporations. As a Center for American Progress writer stated, "Bold policy solutions are needed to tackle corporate concentration and power, empower farmers to negotiate fair prices, and ensure that farmers receive a fair share of the fruits of their labor."[46]

When Industrial-Size Agriculture Began

Starting in the 1970s, and with the encouragement of such politicians as Secretary of Agriculture Earl Butz, who said to US farmers, "Get bigger or get out," we began to see the development of industrial-size agriculture. Farms got larger and farmers' debts greater. Corn and soybeans became popular crops for this new industrial agriculture. Dairy, poultry, beef, and hogs became part of huge industrial-size operations.

Soon mega-agribusiness firms emerged to the point that by 2020 a high percentage of the following agriculture sectors were controlled by as few as four corporations: beef processing: 85 percent; soybean processing, 80 percent; pork processing, 67 percent; and chicken processing, 54 percent.[47] Four international firms control more than 60 percent of global proprietary seed sales.[48]

Food Systems and Rural America

In 2019, writer Austin Frerick interviewed an official of the Iowa Farm Bureau, and from that interview wrote these comments:

> The outlook for rural communities is grim. There are fewer jobs than there were a generation ago and the ones that remain pay lower and lower wages. America's agricultural system is predicated on an extractive model, where more and more of the profits flow to a few. If current trends continue, rural America will soon be owned by a handful of families and corporations who will run their empires remotely with driverless tractors and poorly paid staff. . . . Economic power is more concentrated today than at any other point in American history, and nowhere is this power more apparent than in agriculture. The American food supply chain—from the seeds we plant to the peanut butter in our neighborhood grocery stores—is concentrated in the hands of a few multinational corporations.[49]

Another writer, in 2021, challenged the assessment that the natural order of things in American agriculture was industrial farming and that multinational market domination would continue. Today's agriculture "didn't grow out of consumer demand for fair and competitive markets. Industrial agriculture is propped up by a vulnerable business model that has publicly failed. It owes its entire existence to individual actors—at the US Department of Agriculture (USDA), in finance sectors, in Congress, and certainly at the White House—all who made specific, intentional decisions to promote it [industrial agriculture] as the future of agriculture."[50]

Critics of Industrial-Size Agriculture

Critics of industrial-size agriculture have had difficulty being heard, and if heard at all, dismissed as old-school and not up-to-date with what is happening in the world. Wendell Berry, a Kentucky farmer and author, is one of those who has stood up to industrial agriculture and its ultimate, negative effects on communities, farmers, farm families, and the environment. Berry is a proponent of agrarianism. He writes about a fundamental difference between industrialism and agrarianism.

"I believe that this contest between industrialism and agrarianism defines the most fundamental human difference, for it divides not two nearly opposite concepts of agriculture and land use, but also two nearly opposite ways of understanding ourselves, our fellow creatures and our world." Berry goes on to say, "Because industrialism cannot understand living things except as machines, and can give them no value that is not utilitarian, it conceives of farming as forms of mining; it cannot use the land without abusing it. . . . Industrialism begins with technological invention. But agrarianism begins with givens: land, plants, animals, weather."[51]

I bring back Aldo Leopold, whose land ethic approach challenges industrial-size agriculture, which appears to see money and profits as important above all other considerations.

"Quit thinking about decent land use as solely an economic problem. Examine each question in terms of what is ethically and esthetically right, as well as what is economically expedient. A thing is right when it tends to preserve the integrity, stability and beauty of the biotic community. It is wrong when it tends otherwise."[52]

Author Mark Shepard is critical of where agriculture is today and offers suggestions for improving the situation. He writes, "What is needed are ecosystems that are designed to produce our food, fuel, animal feed, medicine and fibers, and ecosystems that can do so without the use of fossil fuel technology, those that can tolerate extremes of weather and potentially changing climates, and that can thrive without supplemental irrigation from vulnerable and increasingly expensive public utilities." He goes on to say, "We must begin making long-term, permanent change in a world of short-term thinking. In a world impatient for a quick fix, we must continue to make the long, steady progress needed toward a rich, green abundant world, started by planting one tree at time and repeated over and over around the world."[53]

In the following section, I examine two examples of industrial-size agriculture that are becoming ever more common in America today: industrial-size animal agriculture and industrial-size crop production. I look at them from several perspectives—what and where they are, why their numbers are expanding, and the challenges they are creating for the future of agriculture in this country.

X
Industrial-Size Agriculture

Large-Scale Beef-Cattle Feeding Operations

A large animal feeding operation is referred to as a concentrated animal feeding operation (CAFO) by the Environmental Protection Agency. A CAFO has more than one thousand animal units (an animal unit is defined as an animal equivalent of 1,000 pounds live weight and equates to 1,000 head of beef cattle, 700 dairy cows, 2,500 swine weighing more than 55 pounds, 125,000 broiler chickens, or 82,000 laying hens or pullets) confined on site for more than forty-five days during the year. Feed is brought to the animals rather than the animals grazing or otherwise seeking feed in pastures, fields, or on rangeland. Manure is generally collected in huge human-made lagoons. CAFOs are regulated by the Environmental Protection Agency under the Clean Water Act in both the 2003 and 2008 versions of the "CAFO rule."[54]

Feedlots are commonplace in many states. Farmers ship their beef cattle to feedlot operators to fatten them on a high grain diet for several months. Sometimes a feedlot may contain as many as ten thousand beef animals. Consider that each one eats about twenty-five pounds of feed each day, resulting in about nine pounds of manure every day per animal. A continuing problem for operators of CAFOs is what to do with the manure. Sometimes it is spread on nearby farm fields, other times it may be composted and later sold as garden fertilizer. But nonetheless, manure smells, and to the nonfarmer, it smells bad.[55]

A few years ago, I remember traveling on Interstate 80 in Nebraska. We came upon a beef feedlot similar to what I've described here. Even for this farm boy the stench was awful. I thought about the health of the animals as some of them stood in the muck and manure in some feedlot pens, with little room to wander far. We stayed in a motel several miles away from the feedlot and the smell of cattle manure still hung in the air.

Hog CAFOS

The state of Iowa is the leading pork producer in the US. Nearly one-third of the nation's hogs are raised in Iowa.[56] The vast majority of these hogs are found in CAFOs. In Iowa, CAFOs have increased fivefold during the past two decades. In 1990, Iowa had 789 CAFOs, those housing 1,000 or more animals. By 2019, the number had grown to 3,963. Almost all of the growth is from hog-feeding operations.

> Iowa, the top hog-producing state, housed more than 22.7 million hogs in 2017, an increase of 8.5 million since 1992. Large hog CAFOs house a minimum of 2,500 pigs each, and the largest hog CAFO in Iowa houses 24,000 animals. In total, more than 60 percent of the animal waste produced by the largest CAFOs in Iowa comes from hogs. The mountains of animal waste produced by these facilities pose a serious and growing threat to human health, the environment and water resources in the state.[57]

Dairy CAFOs

The Environmental Protection Agency defines a dairy CAFO as one with seven hundred or more cows confined in buildings, and who are not allowed to move outside as their feed is brought to them. My home state of Wisconsin has several dairy CAFO operations—especially in southern and northeastern counties—which are regulated by the Wisconsin Department of Natural Resources.

It is easy to identify these operations. The dairy barns, very different from Wisconsin's historic dairy barns, are steel buildings with canvas curtains on the sides that can be raised and lowered. Inside are dairy cattle in two long rows eating in mangers with a driveway between them. Other buildings include a milking parlor where the cows are milked. Additionally, there are long rows of little white buildings. Each one houses a calf. Usually there is also a tanker truck or two near the buildings used to transport the milk from what may be five or more thousand dairy cows at the farm. And nearby is an open-air lagoon where the manure is stored prior to it being spread on neighboring fields. Dairy CAFOs are required to have a manure lagoon with enough capacity to hold six months' manure. Once the manure spends several months in a lagoon, its smell seems to intensify.

By 2021, Wisconsin had 289 Dairy CAFOS. In that year, the state had 6,804 dairy farms, down from 14,158 as recently as 2007. The number of CAFOs in Wisconsin, 90 percent of which are dairy, has risen each year since 2005.[58]

Thousands of small and mid-size Wisconsin dairy farms have closed as the industrial-size dairy operations have essentially taken over the industry. But this has not occurred without problems, especially problems related to the environment. According to the US Environmental Protection Agency, industrial-size agriculture is the leading contributor of pollutants to lakes, rivers, and reservoirs in the United States. Large animal waste lagoons can also result in significant greenhouse gas emissions—a contributor to climate change.[59]

"With so many animals housed in one place, CAFOs produce volumes of manure every day that somehow must be dealt with. The most common practice is to spread this manure on farm fields near a CAFO in the spring and fall. The manure helps return nutrients to the soil for crop production, but it also has the potential to seep into groundwater and wash into streams and lakes."[60]

One answer to the manure problem presented by dairy CAFOs is an anaerobic digester. "Manure digesters use different combinations of microbes, heat, water and physical agitation to process animal waste. Three main substances come out of the process: methane gas used as a renewable energy source, liquid manure used for fertilizer, and solid manure for composting and cow bedding. Digesters have the added benefit of cutting down on farms' greenhouse gas emissions and livestock's considerable impact on climate change."[61]

Dairy CAFOs have advantages. They generally are more efficient than smaller dairy operations—more milk can be produced per animal at less cost. The farm owner doesn't have to milk cows twice or three times a day, 365 days a year, but can hire workers to do it. A CAFO can be an important employer in the community. The CAFO operator can concentrate on one agricultural endeavor, in this case, dairy, without having to become proficient in several, which was the case for the diversified farm owner. On diversified farms, not only did the farmer have to be proficient in working with several different animal species such as dairy cows, hogs, and chickens, but he or she also needed to be proficient in growing several different kinds of agronomic crops such as alfalfa, clovers, and pasture grasses along with corn, oats, and perhaps soybeans. Some CAFOs do grow much of their own feed on land they rent or own.

The Future for Dairy CAFOS

A few years ago, I was giving a talk about agricultural history at a museum in Green Bay, Wisconsin. In my talk I shared information about the development of dairy CAFOs, or said another way, industrial-size dairy farming. I always save time at the end of my talks for questions from the audience. A fellow near the back held up his hand. "Could I talk to you later?" he asked.

"Sure," I answered, wondering what he had on his mind. When this happens, and it has happened before, the person usually wants to tell me how much they disagreed with what I had to say, or they try to correct me on some fact I shared. So, I didn't look forward to this encounter.

When I came out into the lobby, this fellow was waiting for me. He appeared to be in his mid-thirties.

"Let's find a place to sit down," he said. His voice had a friendly tone to it. Once seated, he began, "Thank you for sharing information about Wisconsin's agricultural history, especially the history of dairy farming in Wisconsin."

"You are welcome," I said, wondering where this discussion was going.

"I'm a large animal veterinarian," he began. "Most of my clients are farms with at least one thousand milk cows. I'm convinced that this kind of dairy farming is not sustainable. There will come a time when they will fail."

I couldn't believe what I was hearing. These large dairy CAFOs were his clients. He went on to tell me about the pollution problems, especially groundwater pollution, that some of these farms were causing. He then shared several personal experiences he had that convinced him these huge dairy farms would ultimately fail—especially if they continued their operations in a manner similar to what they were doing today.

As I thought about my research and my personal experience with CAFO operations, I began considering what is driving the CAFO movement in this country. Here are a few beliefs that I think drive the CAFO movement:

- Bigger is better.
- A CAFO operation can be compared to a factory where inputs and outputs are carefully considered.
- For animal agriculture to survive, large, efficient operations are necessary.
- A farm animal in some ways can be viewed as a production machine.
- Economic considerations are more important than environmental ones.
- Land is of value only for its ability to support a CAFO operation and provide a place to spread the manure.

Industrial-Size Crop Farming

As animal agriculture has followed the mantra of "bigger is better," so has crop farming. This is especially true for two major commodity crops: corn and soybeans. Following World War II, when agriculture began to change and change dramatically, industrial-size monoculture crop farming (planting the same crop on the same large-acre field year after year) emerged. There are some advantages to this type of farming. As a specialized form of agriculture, it can be efficient with a lower cost of production. Monoculture crops can be easier to manage. A farmer needs only to learn the details of growing one crop.

Yet the disadvantages of monoculture cropping are several. Because of the lack of biological and genetic diversity, weeds and insect pests can spread faster in monocultures. There are no natural defenses or barriers of co-occurring, naturally repelling plants to stop pest infestations. Because of this loss of diversity and the natural resistance provided by other plants and organisms, farmers end up added increasing amounts of fertilizers, pesticides, and water to their monoculture fields.[62]

In addition, monoculture farming degrades the soil. With heavy applications of fertilizer and weed killers, soil organisms such as bacteria, fungi, and earthworms, which normally break down organic matter, distribute nutrients, and remove parasites or other harmful organisms, disappear.

"No matter how much water and fertilizers are applied, these soils are not able to take them in and utilize them efficiently for crop growth. The result is increased runoff of concentrated nutrients that pollutes surrounding areas, while cultivated lands yield less and less. . . . Excessive fertilizers and pesticides not only change the chemistry of surface water, they also leak into the groundwater and drinking wells."[63]

Pesticides that control pests on monoculture crops also kill bees and other pollinators. Researchers at the University of Wisconsin–Madison studied the decline of bumblebees (an important pollinator species). "As farmers cultivated more land and began to grow fewer crops over the past 150 years, most bumble bee species became rarer in Midwestern states."[64] As the bees disappear, yields of crops that require pollination decline. Examples include fruits crops, such as cranberries and apples, but also vine crops such as cucumbers, squash, pumpkins, and many others.

The majority of the corn and soybeans planted in monoculture industrial-size farming in the United States are GMO (genetically modified organism) crops. This means that the genetics of the plant have been modified so that it is resistant to insecticides or herbicides (such as Roundup), or often both. GMO crops were first introduced in the US in the mid-1990s.[65] Strong arguments have arisen about the advantages and disadvantages of growing GMO crops.

GMO crops have several advantages—a major one is that weed and pest control require less effort—but they also have disadvantages. Pesticide and herbicide resistance can develop. Even when weed and insect pests are chemically controlled, some always survive. These survivors pass on their resistance traits to the pests that follow them. With each crop planted, more survival pests—sometimes known as superweeds—appear. And heavier applications of herbicides do not usually control them.[66]

Just as I attempted to identify some of the beliefs that appear to undergird industrial-size animal agriculture, I have tried to do the same with monoculture crop farming:

- To survive, a crop farmer must have a large operation.
- Economics is more important than the environment.
- Research and technology will eventually solve any problems that emerge, such as control of superweeds and new insect pests.

- Concern for the environment is somebody else's problem.

XI
Assuring a Future Food Supply

Of the several employment sectors available in rural communities, agriculture is unique. There are a few exceptions to this statement as we see an increasing number of food production units operating in urban areas. Food production on farms, however, will continue into the future. And farming, with appropriate adjustments and changes, will continue to be a major economic engine in rural communities, as well as being the major food producer in the country. But tomorrow's agriculture will not be a reflection of what is happening today. Dramatic changes must be made in the agriculture of the future. And it is the responsibility of everyone to help make sure these changes occur.

It is easy to assume—especially if you are someone who lives in an urban area—that the problems facing agriculture are not your problems. You may believe that it is up to agricultural colleges and universities to figure out the answers. Or that they are the problems of farmers, agricultural organizations, including farm organizations and the vast array of food-processing and distribution organizations, and local and federal policymakers. They are the problems of environmental organizations and activists. All true. But agriculture's problems are everyone's problems, no matter where you live and what you do—because we all have to eat.

Let's began by looking at some of the problems and challenges that agriculture faces today. First, industrial-size farming and the consolidation of power in a handful of food-processing and distribution corporations are major problems. Some writers see the present situation as grim. "Vertical, horizontal, and backward integration at local and global levels has trapped farmers in suffocating contracts, exploited workers, and degraded farmland with extensive monocropping and chemical inputs."[67]

Reflecting on the thinking of Aldo Leopold and the Native American's Seventh Generation philosophy, these writers continued:

> The earth is not a passive entity to be exploited for maximum productivity. As Indigenous communities have known for centuries, the earth is a naturally regenerative body that can be cultivated to remain ecologically productive for thousands of years. Farming and land stewardship go hand-in-hand, farmers are necessarily invested in the health of the ecological systems that sustain them. In pushing a narrative of industrial necessity, Big Ag has left a mess of environmental, social, and economic destruction in its wake.[68]

What can be done about this situation, which appears to be expanding? In place of the present situation with a few huge corporations in charge of our food system, the agriculture sector needs to advocate and help create a competitive food system with many interdependent producers, processors, and distributors.

> Within this system, we would see more farmers selling directly to consumers, food hubs distributing locally-produced products, independent processing plants serving farmers in their regions, community supported agriculture

> networks ushering produce from farm to table, and an abundance of local options in retail grocery stores. A democratized system would give each player along the supply chain—farmers, workers, processors, consumers—the ability to make the best decisions for their businesses, their communities, and their families.[69]

Alternatives to Industrial-Size Agriculture

Sustainable Agriculture

Sustainable agriculture is defined as

> an integrated system of plant and animal production having a site-specific application that will, over the long term, satisfy human food and fiber needs; enhance environmental quality and the natural resource base upon which the agricultural economy depends; make the most efficient use of nonrenewable resources and on-farm resources and integrate, where appropriate, natural biological cycles and controls; sustain the economic viability of farm

> operations; and enhance the quality of life for farmers and society as a whole.[70]

The sustainable agriculture movement provides an alternative to industrial-size farming, while at the same time offering a response to the growing need for environmentally sound agricultural practices on farms of any size.

Sustainable agricultural practices generally include integrating livestock and crops, rotating crops to ensure healthy soil as well as pest control, and planting cover crops, especially in the nongrowing season on land that would otherwise be bare. These practices help build soil health, prevent soil erosion, keep weeds in check, and replenish soil nutrients. Practicing reduced soil tillage such as planting seeds in undisturbed soil helps reduce erosion and improves soil health. It also includes using a minimal amount of chemical pesticides, following integrated pest management practice, and using mechanical and biological controls for crop pests.[71]

One excellent example of a practical application of sustainable agriculture is Superior Fresh, located in Hixton, Wisconsin. It is a family farm that produces both certified-organic leafy greens as well as well as organic seafood. In its one-acre fish house, it produces about 160,000 pounds of Atlantic salmon and steelhead each year. Here is how they describe their operation: "Our responsibility is to provide people with the safest, most nutritious food possible, while mindfully restoring the land around us."[72]

Organic Farming

Sometimes we need to look back in order to look forward. Many of us, without calling it that, practiced organic farming in the early days of our lives. On the home farm when I was a kid, we practiced crop rotation, used manure for fertilization, used mechanical weed control (cultivators and hoes), used no GMO seed (there wasn't such a thing), and planted cover crops on fields that would otherwise be open during the nongrowing season. All of these practices and more have been formalized for organic farming.

Today, organic farming is defined as "an agricultural system that uses ecologically based pest controls and biological fertilizers, derived largely from animal and plant wastes and nitrogen-fixing cover crops."[73] Organic farming not only produces nutritious food, it also is concerned for the preservation of natural resources. In the United States, the standards for organic certification are set by the US Department of Agriculture, with the backing of the National Organic Standards Board, a group of volunteers that includes organic farmers, those who sell organic products, environmentalists, scientists, and other advocates of organic farming.

Standards for Organic Farming

- Land must be free of all prohibited substances for at least three years before crops can be certified organic.
- No genetically modified organisms (GMOs)
- No use of synthetic fertilizers or pesticides
- Certification is available to farmers and food processors who sell more than $5,000 in organic products a year.
- Crops, livestock, processed products, and wild-harvested plants can be certified.
- Farms and processors must undergo regular inspections to ensure compliance with National Organic Program standards.[74]

Organic farming has several shortcomings. Yields of organic farming are generally lower than those of conventional farming, thus it takes more effort to produce the same amount of food in organic farming as compared to conventional agriculture. This means that prices for organic products are higher than crops grown conventionally. With lower yields of organic crops, more farmland is needed to produce the same amount of food that would be produced by conventional farming, thus increasing human pressure on ecosystems.[75]

Small Acreage Farming

Small acreage farming takes several forms. I define small acreage as 160 acres or fewer. Some of these small acreage farmers farm organically, some operate as farm-to-table farms (see below) and have contracts with local restaurants and grocery stores. Some operate as part of cooperatives, with several small acreage farmers working together and operating a retail store in a nearby village where their produce is sold. In my novel *Settlers Valley*, I write about a group of veterans with both physical and psychological challenges who operate small acreage farms as well as run a retail store where their produce is sold. An additional dimension to my story of these veterans, who are farming what is essentially “worn out” land, is that they are healing personally as they farm. I write that as these veterans heal themselves, they also heal the land.[76]

Community-Supported Agriculture

Some small acreage farmers are part of the community-supported agriculture (CSA) movement. An urban family purchases shares directly from the farmer, depending on the membership agreement, and the farmer delivers farm shares—boxes or bags of produce—weekly or biweekly throughout the growing season. A typical share can include lettuce, strawberries, cheeses, eggs, flowers, meats, and even preserves. The contents of the box vary as the season progresses.

Home Gardens

During World War II, growing a garden in your urban backyard became popular. These gardens, known as Victory Gardens, produced literally tons of fresh vegetables for urban dwellers who often found it difficult to find fresh produce during the war.

The idea has caught on once more, with people interested in knowing the source of their food and answering the question by growing their own. A home garden can be as small as a large flowerpot where you might grow a tomato plant or some lettuce. Or it could be a raised garden of the type I have grown in my backyard. My raised garden was four feet by eight feet and was surrounded by a fence to keep out the critters, especially the rabbits who in the past have enjoyed my lettuce crop more than I have. In my raised garden I have grown lettuce, climbing tomatoes, climbing cucumbers, climbing beans, and a few flowers. My wife and I have enjoyed fresh vegetables from this little garden all summer, from the lettuce in spring to the climbing beans that were still bearing until the first frost in early October. In fact, the climbing beans—and all the climbing crops attached to the fence—grew to six feet tall. All one needs for a backyard garden is six to eight hours of sunlight plus a little weeding and occasional watering. And I should add, taking care of the little garden was an avocational joy—especially if you enjoy growing things and marvel at what can come from a little seed.

Additionally, we grow a large, family-size garden at our farm. There we grow potatoes; vining crops such as squash, pumpkins, and zucchini; several rows of sweet corn; forty to fifty tomato plants; carrots; beets; sunflowers; kale; peas; radishes; and more. Three families have fresh produce from our farm garden, which my son and daughter-in-law manage. They do the planting, weeding, watering, and harvesting. We have had this garden since 1967, the year after we bought our farm.

Farm-to-Table Movement

A rapidly growing and widely accepted alternative to industrial-size food production is the farm-to-table movement. Farmers, generally small acreage or organic farmers, sell their produce directly to nearby restaurants where customers enjoy the fresh food as well as appreciate the decreased environmental impact of food grown nearby. Advantages of farm to table are many. Consumers get to enjoy peak freshness in their foods, as they likely have only traveled a short distance from the farm to the consumers' table. The approach is environmentally friendly as hundreds of gallons of fuel are saved if the food is available nearby rather than traveling great distance, thus air pollution is largely avoided from the burning of fossil fuels. This approach supports local farms while also supporting local restaurants and grocery stores. Farm to table negates consumers' reliance on large corporate food companies for consistent, high quality, fresh food. Farm to table also can help urban people learn about the food resources in their community.

One interesting example of a farm-to-table operation is one that operates at a retirement community, The Villages, in central Florida. The Villages' farm-to-table initiative, called "Villages Grown," includes full production at its vertical greenhouse for microgreens, as well as three new types of lettuce. Villages Grown operates a retail store at Brownwood Paddock Square as well as a mobile market.

The Villages Grown is both a grower and a reseller of agricultural products, as is true of other farm-to-table producers. It partners with some seventeen local farms in addition to the crops grown in its greenhouses located on forty-five acres of land it owns. Growers use vertical hydroponics, using stacked towers to grow its crops. Their system allows them to prevent pests and diseases as well as extreme weather from negatively affecting their crops. Villages Grown grows twelve different kinds of microgreens, including kale, radishes, beets, chard, kohlrabi, and broccoli.[77]

Vertical Hydroponic Farming

Different from traditional farming where crops are grown in soil and are heavily dependent on climate and weather, vertical hydroponic farming crops are grown indoors, and no soil is needed. Crops are planted in trays or towers and fed with water treated with the specific nutrients required for each plant type. Within the growing house with its vertical walls, light, heat, and humidity conditions are carefully controlled for each crop. Hydroponic gardening uses a sterile growing medium, thus weed, disease, and pest problems can be minimized. Crops such as corn, squash, pumpkins, and soybeans and fruit crops such as oranges, apples, and cherries do not lend themselves to this type of agriculture. Common crops for vertical hydroponic farming include lettuce, spinach, strawberries, bell peppers, and herbs.

With this system these crops can be grown year-round, and they require far less space than needed for traditional farming. "The complete lack of soil and closely controlled microclimates for each plant answer concerns relating to climate change and soil degradation. In fact, an acre and a half of vertical hydroponic farming can produce 500 tons of healthy, leafy greens spread over 11–13 harvests each year."[78]

Aeroponic Farming

Aeroponic farming, similar to hydroponic farming, allows farmers to avoid risks related to weather and pests. Because crops are grown stacked inside a building, only minimal acreage is necessary. Instead of soil, aeroponic farming crops are suspended in a fine mist of water and nutrients. A pump, connected to a timer, sprays an aerated nutrient solution on the roots periodically. This process saves 95 percent of the water compared to farming with soil and 40 percent of the water used in hydroponic farming where the crop roots are submerged in water. Aeroponic farming also controls lighting, making sure that plants get the light they need, and only the light they need, which improves efficiency and reduces energy costs. Common crops grown include lettuce, strawberries, cherry tomatoes, mint, and basil.[79]

Aquaculture

Aquaculture includes two basic types: marine and freshwater. Marine seafood, which is generally grown in confined areas in an ocean include shrimp, clams, oysters, and mussels. Bass, catfish, tilapia, trout, and salmon are grown in freshwater ponds.

Commonly referred to as fish farming, aquaculture has a long history. It is believed that the Chinese were practicing aquaculture as early as 2,000 BC. The practice soon spread to other countries in Asia and Europe, and later to the United States. Popular fish farms in the United States produce salmon, trout, tilapia, catfish, and cod.

Alternatives to Industrial-Size Animal Operations

Sustainable agriculture approaches, which integrate animals and crop growing, are a strong and environmentally friendly alternative to industrial-size animal operations. (See earlier for details on the sustainable agriculture approach, which is being advocated by many agricultural colleges and other groups). Nonetheless, there are other alternatives to industrial-size animal operations, where no meat and sometimes no animal products are consumed.

Vegetarianism

An extreme answer to industrial-size animal farming is for people to quit eating meat and become vegetarians.

There are several subcategories of vegetarians. Ovo-lactarians eat dairy products and eggs but no meat. Lactarians eat dairy products but abstain from meat and eggs. Some people include fish in their diet but still consider themselves vegetarians. Vegans are the strictest subcategory of the vegetarian movement, abstaining from all animal-based products. Strict vegans do not eat honey or wear leather or wool.[80] The number of vegetarians is increasing in this country and around the world. One measure is the increased retail sales of nut-based milk, plant-based meats, seafood, and other such products.

It's not just vegetarians who are eating plant-based "meats," but many meat eaters are trying them as well. In a survey of 30,700 people, 71 percent reported they had heard of plant-based meat substitutes offered at fast food restaurants. Fifty-four percent had tried it—and of these, 72 percent identified themselves as meat eaters.[81]

> Over the past few years, the fake-meat phenomenon has gone from being a highly speculative curiosity to an in-demand food item, available not just in grocery stores but also dished out over the counters of popular fast-food chains

> across the country. . . . From burgers and sandwiches to tacos and pizzas, a wide range of meatless alternatives are available to customers today, many of whom are opting for them for health, environmental, or ethical reasons.[82]

By 2020, firms were producing plant-based food in these categories: whole-cut meat, seafood such as plant-based crab cakes, fish cakes and fish burgers, chicken, beef, pork, cheese, eggs, yogurt, ice cream, and milk.[83]

Dairy Alternatives

People in the United States are not drinking as much cow's milk as they once did. For example, in 1975, the average American drank 247 pounds of milk (115 quarts) per year. By 1990, milk consumed per person per year declined to 220 pounds (about 102 quarts). In 2000, the number dropped to 197 pounds (92 quarts), and in 2020 the number reached 141 pounds or (66 quarts) per person per year.[84] Several reasons are offered for the decline. One reason is that a sizable number of people are lactose intolerant and suffer health problems from consuming dairy cow products.[85] The vegetarian movement has convinced some people that they would be healthier if they consumed "milk" that comes from nondairy sources. Visit any grocery store these days and you'll see a considerable display of nondairy "milk." The most popular ones are almond milk, soy milk, coconut milk, and cashew milk. Examples of nondairy cheeses include Daiya Mozzarella Style Shreds, Vegan Gourmet Cheddar, and Teese Cheddar Vegan Cheese.[86]

Dairy Goats

Goat milk is increasing in popularity as an alternative to cow's milk, especially for people who are lactose intolerant and must avoid cow's milk. Goat milk has more protein than cow's milk, and is loaded with calcium and other minerals such as magnesium and potassium when compared to the various nut "milks." Goat milk is creamy, with a consistency similar to cow's milk. It has small, well-emulsified fat globules and cream that will stay in suspension for a longer period of time than cow's milk, thus goat milk does not need to be homogenized.[87] Goat cheese has also become a popular specialty cheese among cheese lovers.

Laboratory-Produced Dairy Products

In addition to plant-based milk alternatives, a new type of milk is being researched. The company producing it is called Perfect Day. They said this in a promotional piece: "Finally, an option for dairy lovers and plant-based fans alike. We've invented the world's first milk proteins made without animals, so you can enjoy the taste, texture, and nutrition of traditional dairy, produced sustainably and without the downsides of factory farming, lactose, hormones, or antibiotics." Perfect Day's products are different from plant-based products, such as almond, coconut, and soy milk. They result from creating proteins that are genetically identical to animal proteins. The company's protein product is made by altering sections of the DNA sequence of food-grade yeast such that the microorganisms, once fed with certain nutrients, produce several key proteins found in milk.[88]

Ryan Pandya, one of the founders of Perfect Day, says, "Developing this science is about more than creating a tasty vegan-friendly [milk product]. Cutting down on the number of cows is a way to make the food industry more sustainable. These animal-free 'milk' products use 98 percent less water and 65 percent less energy to create [milk products] than ones that use dairy cows."[89]

As evidence continues to mount about the environmental concerns of industrial-size agriculture, including air and water pollution concerns, alternatives, such as those mentioned here, will become more popular. Developing opposition to the industrial model for agriculture, and the tendency to look at agriculture as merely an industry without considering its relationship to nature and the land, will result in a new way of looking at agriculture and its future.

XII
Future Rural Communities

Historically, rural America was made up of villages, many of them with populations of less than one thousand people, and open areas, much of it devoted to farming. These rural communities depended on agriculture, manufacturing, and mining for their economic well-being. The service sector, including health care and food services, were also important employers in rural communities and will continue so into the future.

As I explained earlier, these villages and the farmers who lived and farmed in the open areas were closely linked to each other. The farmers depended on the villages as a market for their products, and the villages depended on the farmers for purchasing the basics for their farming operations such as feed, hardware, and other farming supplies, as well as providing basics such as sugar, coffee, flour, and clothing for the farm family. With the revolution in farming that occurred after World War II, and the more recent emergence of industrial-size farms, this dependence of village to farmer and farmer to village has largely disappeared. With all of these changes in mind, let's examine the future for rural communities in America.

In 2022, there remain thousands of rural villages, from those with a few hundred to those that have a few thousand in population. The vast majority of US land is considered rural. As discussed earlier, rural land use has included agricultural, residential, industrial, mining, and recreational. Each of these uses, depending on how they are applied, has an effect on the environment. Or to look at it from Leopold's perspective, how they, to a lesser or greater extent, violated his idea of a land ethic.

Over several decades many people have come to believe that rural America's decline is inevitable. "The popular understanding of this decline follows a familiar script: previous generations age and die off, new generations move to tech-hub cities across the country, and the declining monochromatic economies and demographics left in their wake provide little motivation for residents to move back or new ones to consider their future there."[90]

Since World War II, the rural population has decreased. Rural young people fled to the urban centers. But a new trend is developing. In 2018, the Pew Research Center found that more urban (30 percent) and suburban (35 percent) residents are interested in moving to a rural community than rural residents (20 percent) are interested in moving to an urban community.[91]

At this writing the numbers are not yet available, but the COVID-19 epidemic (in the US from 2020 to 2023) has had an interesting effect on urban-rural migration. When people discovered that they could work at home using their computers and the internet, they discovered they had options as to where they might live. Some of them were forced to work at home, as their workplaces closed to avoid COVID spread; others were given a choice to do so, and they discovered that they liked it. The next thing they discovered was that they didn't need to live in a cramped, expensive apartment in a big city but could instead move to a rural community and discover a different lifestyle while continuing to work from their new homes.

Unfortunately, many rural communities were not ready for an influx of work-at-home professional people wanting to move to their rural community. These new residents had needs that rural people had not yet considered of high priority—broadband internet availability topped the list. It would be impossible for these new urban arrivals to carry on their work without a modern broadband system. Not far behind in importance was the need for a robust health system with sufficient health-care professionals, hospitals, and medical clinics. First-rate educational opportunities also ranked high for urban workers with young families when they considered a move to a rural area. During the past several decades, many rural areas have been losing population resulting in declining school enrollment and school closings. Adequate funding of schools is a continuing problem. Because rural families are more scattered, and the schools are some distance away from them, expenses for transporting children to and from school, and to after-school activities such as athletic events, can be considerable. A related consideration is that rural communities generally have a larger percentage of older people than an urban population.

Cultural opportunities, such as modern community libraries, parks, good restaurants, musical and theater events, museums, and other cultural activities taken for granted in urban areas, are often missing or inadequate in rural communities. Many small, rural villages have become food deserts. Some do not even have a grocery store. It seems a contradiction. All around the village there may be fields of soybeans and corn—thousands of acres of them—and yet in the village, residents may have to drive miles to buy groceries. About five million people in rural areas have to travel ten miles or more to buy groceries, according to the Department of Agriculture.[92] Many rural villages from Florida to Montana are fighting back by starting their own community-run markets. More about this to follow.

Some rural villages, especially those with outdoor recreational attractions that were once farm service centers, have changed, with an emphasis on serving the tourists visiting the area. Many of these changes are the same ones that would invite urban people to live in the community—adequate health services, cultural opportunities, and so on.

An excellent example of a farm service town that has transformed itself into something much more is Viroqua, Wisconsin. Viroqua, with a 2020 population of 4,418, is the county seat for Vernon County. It has many benefits of a large city. It is located in the "driftless" or "unglaciated" region of southwestern Wisconsin. It is a region that the great glaciers didn't reach. Left behind were spectacularly beautiful ridges, bluffs, valleys, and rolling hills. Viroqua features numerous natural resources and outdoor activities, such as hunting, fishing—excellent trout fishing—hiking, camping, canoeing, cross-country skiing, and swimming. It is a town with a rich history, a thriving arts and entertainment scene, and a variety of small businesses. The city has a food cooperative and a weekly farmers' market stocked with locally grown fresh produce. Viroqua includes an array of education and health-care options, including a hospital. The city is in the center of one of the largest organic farming regions in the United States. The region's many organic farms provide fresh produce to local restaurants.

The town traces its history to 1846, when Moses Decker arrived in the area. Several of Viroqua's original buildings trace back to Decker's work. The downtown district has been named to the Wisconsin and National Register of Historic Places. The local theater has been restored and is an excellent place for plays and musical performances. For those interested in shopping, many small shops offer a variety of items including locally created artwork. Viroqua's several restaurants offer everything from organic food to supper clubs. Likewise, a variety of lodging opportunities are available from motels to historic inns. Entertainment includes an eighteen-hole golf course, the Driftless Music Festival in July, and in August, Wild West Days, which takes Viroqua back to an 1880s boom town. And not to forget the Vernon County Fair in September, which features five days of animals, exhibits, carnival rides, a tractor pull, and harness racing.

> The residents of Viroqua are widely diverse in their lifestyles, yet find common ground in their ability to work together to accomplish great things. Emphasizing collaborative leadership and preservation of our history, the community and Viroqua Chamber Main Street continue to involve residents in vital and meaningful ways. Volunteers of all ages are engaged in ongoing projects which contribute to the sense of well-being for all. The Viroqua Chamber Main Street is dedicated to enhance the quality of life of the Viroqua community through leadership, facilitation and collaboration.[93]

Rural villages have a potential for a bright future. But that future will not be a repeat of what these villages experienced in the past, when they mostly served as partners with the farmers who farmed in nearby lands. If agriculture gets past the time when it believes that its future is industrial-size operations, with only a handful of people living on the land, and moves toward the alternatives outlined in this book, some of this interdependence of villages and farmers will return. We will see more farm-to-table restaurants, fresh food grocery stores, and farmers' markets where fresh produce can be purchased. Many villagers will also have their own gardens to produce fresh vegetables. The villages and small cities of the future must also realize that their future must include urbanites who see these rural places as destinations to visit for their recreational and outdoor experiences.

Increasingly, urban people are leaving large urban centers and moving to rural communities to live and work—albeit via their computers and broadband availability. These rural villages and small cities of the future will have first-rate educational opportunities and the most up-to-date medical facilities as well as a variety of cultural opportunities, including modern libraries, interesting museums, and drama and music opportunities.

I see a promising future for rural communities, but it entails some creative thinking that involves both the villages and small cities, as well as the vast acres of farmland that surround many of these towns. The future of rural communities also means creative thinking of urbanites, as city and country learn to work together.

XIII
Conclusion

The United States depends on strong rural communities, and especially on a vibrant agriculture that focuses first and foremost on the land. The land must be nurtured and respected. We must consider the perspective offered by Aldo Leopold and Native Americans—land is so much more than dirt—it has life, it is complicated, and not only deserves our respect but requires that our decisions about it be done in an ethical way. Economics are important in making decisions about land—but economic decisions too often are, and must no longer be, the most important driving force. The Native American philosophy of seven generations proclaims, in addition to a basic concern for the land, economics, physical, social, and spiritual concerns must be considered as well.

"We are all people of the land, no matter where we were born and where we live today. No matter if we are country dwellers or city folk." In my many presentations, I have often said that. In 2015, I published a book called *Whispers and Shadows.* In that book I included a chapter titled "Listening to the Land." I wrote, "In our hurry, hurry lives, the idea of taking time to listen to the land sounds not only a bit strange, and impossible, but also like a task easily set aside in favor of demands deemed more important. Yet listening to the land might be one of the most important things we'll do in our lives, whether we make our living from the land or simply want to continue having a ready supply of food on our tables."[94]

One way of listening to the land is to search out its stories. When my wife and I bought our farm in 1966, I immediately began searching for its stories, beginning with the glacier that formed the land many thousands of years ago. I then found the stories of Native Americans who traveled over and lived on these lands for thousands of years. I followed this by searching for the early white settlers—Thomas Stewart, a Civil War veteran, was the first. He homesteaded our farm in 1867. The stories made the land come alive, gave it a unique personality, made it a special place. I firmly believe that the more we know about something, whether it is land, an old building, or something else in our lives, the more we are likely to take care of it.[95]

Robin Wall Kimmerer, an enrolled member of the Citizen Potawatomi Nation, takes a related perspective for listening to the land when she writes, "The land knows you, even when you are lost."[96]

A second concern that agriculture must face is moving beyond a dependence on the industrial model to guide its way, as it has for the past several decades. John Ikerd, professor emeritus of agricultural economics, University of Missouri, wrote the following about this fundamental change that must occur in our society:

> I believe humanity is in the midst of a great transformation, moving out of the industrial era and into a new fundamentally different era of human progress. . . . It is being ushered in by the sustainability movement, which includes sustainable development, sustainable living, sustainable agriculture, sustainable fisheries, sustainable forestry and sustainable any other aspect of our economy or society. . . . The great transformation is being driven by the growing realization that the . . . economic development paradigm, which currently dominates all indus-

> trial economies, quite simply is not sustainable over the long run.[97]

I believe that once a new agriculture appears, following the fundamentals outlined here, the negative view some urbanites have toward rural American will go away. This negative view of rural America has an interesting root. Norman Wirzba, professor of Christian theology at Duke Divinity School, wrote this: "Without condemning urban life per se, it would be naïve to overlook or ignore the cultural significance of mass migrations from the country to the city, for in the extinction of agrarian practices we witness not only a loss of a way of life, but the cultural sensibility that understood intimately and concretely the human bond with the earth."[98]

Vibrant rural communities with a forward-looking "new agriculture" are essential to the future of this country. They offer employment, recreational opportunities, a source of water and energy, and above all provide food. Kathy Cramer, Author of *The Politics of Resentment*, asks this challenging question: "What would it take for urban people to recognize that they are better off when rural communities prosper?"[99]

Providing food for the country was always and continues to be a vital role for many rural communities. For these communities to continue their important role into the future, several changes must occur in agriculture. Historically, farming has been a mainstay for many rural communities in America. For years in the past and including the present, farms provided food and fiber for the nation. During the past several decades, especially since about 1970, agriculture lost its way as it turned to industrial-size operations. Only a handful of massive producers, processors, and distributors are in charge of much of agriculture. The small- and medium-size family farms have struggled and mostly disappeared in America. While farming was once a way of life, today it has tried to become an industrial-like business.

For rural communities to once more gain the importance they previously had, major changes must occur in agriculture, at all levels, from production to processing and distribution. It is not sustainable for agriculture to continue on the track it has taken. Industrial-size agriculture is not environmentally friendly. It is more concerned with profits than with the environment. It is mostly ignoring the land upon which its future rests.

New Farmers

An ever-increasing number of small- and medium-size farmers are leaving the land. This trend must be reversed if we are to again see thriving rural communities. One way to encourage people to take up farming is for retired and present-day family farmers to share their stories of farm life. What a typical day on a family farm is like. The values that emerge from working the land, including developing an appreciation that it is much more than dirt. Realizing that caring for and respecting land is one of the most important things a person can do. Learning to appreciate the importance of the simple things, such as the joy of watching a sunset, and appreciating the wonders of nature that are all around you.

I learned all of this from my years working on our farm. As a kid, I learned the power of curiosity, which has served me well over the years. When I encountered something new, something I hadn't seen before,

I wanted to know something about it. I learned the importance of slowing down. "What's your hurry?" are my father's words still ringing in my head. I learned to appreciate what I was doing, taking time to learn from that.

Writing down stories about farming and farm life is a powerful way to share what living on the land is all about. In a talk I recently gave, I said this about stories:

> A story is more than words. It can take people to places where they have never been. A story can bring people to life, even though they have been long dead. A story can evoke feelings about people and events that have long been forgotten, bringing tears and laughter, sometimes in the same paragraph. Stories are what makes us human. Our stories make us different from one another, yet tie us all together. People have stories, but so do buildings, cities, villages, farms, trees and prairies, rivers, and lakes. And of course, the land. The land has a story to tell, and telling that story can help others learn and appreciate what we as farmers have experienced.

The story is a powerful way for farm people to help others become acquainted with their lives as farmers, and what life on the farm is all about. For further information about stories and how to write them, see my book *Telling Your Story: Preserve Your History through Storytelling.*[100] One category of stories that many urban people find interesting is what I call the "hidden knowledge" that farm people have: what farm people have learned from their parents, which was passed on from generation to generation. Included are such things as how to grow potatoes from the time they are planted until they are harvested. How to cook and bake on a wood-burning stove, how to tell when cut hay is ready to store in the barn, knowing when it is "the right time" to plant corn on a piece of ground, how to fell a tree with an ax and crosscut saw, and much more. The technical name for this type of knowing is "indigenous knowledge." It is defined as follows: "The knowledge used by local people to make a living in a particular environment. This knowledge is built up by a group of people through generations of living in close contact with nature."[101]

My father, a lifelong farmer, had a fifth-grade formal education, yet he knew more about farming and the land than many people with advanced degrees. He could predict rain by watching cloud formations. He could tell when to plant a crop by holding his hand on the ground. He knew when a grain crop was ready for harvesting by its color. He could tell when corn was ready for harvesting by feeling the corn cobs. And much more. As my father would say, "There is book learning, but there is so much more to be learned that is not in books."

There is much to be learned when farming. It mostly occurs when one doesn't realize it is happening. Learning to enjoy sunsets, working with animals, watching plants grow, enjoying the seasons' changes, plus all the skills needed for fixing a broken fence or knowing how to pile hay bales so they don't fall over.

More people should be encouraged to go into farming and receive help to do so. Colleges of agriculture and their various extension and outreach programs should have special educational programs for these new farmers. Federal and state policies should be changed so that bigger will not necessarily be better.

(You may find the book, *Small Is Beautiful* by E. F. Schumacher interesting).[102]

These new farmers should have the opportunity to explore several approaches to farming, including organic farming, small acreage farming, farm-to-table farming, and community-supported agriculture, all following a sustainable model. Some might also try aquaculture, as well as hydroponic and aeroponic farming. The result should be a competitive system with independent producers, processors, and distributors. These alternative approaches to farming are discussed in some depth earlier.

These new farmers will come from several different backgrounds. One category includes former business and professional people who are tired and disappointed with their careers and want to do something different. Many of them want to raise their children in the country, and they take up some form of farming. Some of them will also continue their former careers in a limited way, following work-at-home strategies.

Military veterans, some of them severely psychologically and physically injured, should be assisted in becoming farmers if they are interested. There is research to show that working the land can have health benefits, especially for those who suffer from such maladies as post-traumatic stress disorder.[103] As I mentioned earlier, in my novel *Settlers Valley,* I tell the story of several physically and mentally injured veterans taking up small acreage farming and showing the benefits they gain from doing so.[104]

Another group is grandchildren of former farmers, who remember spending summers on their grandparents' farm, and how they had enjoyed the experience. Young people who have recently graduated from college and are strongly committed to environmental approaches that include caring for the land is yet another category of potential farmers. As Kathy Cramer points out, "I'm surprised how many young people are looking for job situations that may not provide a lot of money, but provide more balance in their lives. They want a work life that is more humane."[105]

Growing in size is a group of women who have taken up farming as a career change, or as a first career. The US Department of Agriculture's *2017 Census of Agriculture* indicates that 36 percent of US farmers are women and 56 percent of all farms have at least one female decision maker. "According to the USDA, farms with female producers making decisions tend to be smaller than average in both acres and value of production. Women farmers are most heavily engaged in day-to-day farm and ranch decisions, along with recordkeeping and financial management."[106] Women in agriculture are a growing, important segment for a "New Agriculture," which includes their concern for sustainability and caring for the land.

The goal should be to return more people to the land, in a variety of farmer roles. Another goal is the production of food near where it is consumed so that it does not need to be transported hundreds of miles from producer to consumer.

Rural Villages and Cities of the Future

Changes must occur within rural villages and cities as well. From the earliest pioneer days in this country, villages helped tie rural communities together. As mentioned earlier, when the majority of the people in the country were farmers, the villages provided these farmers with the basic necessities of life, such as clothing and food basics such as sugar, salt, and flour, plus offering a variety of cultural and recreational opportunities. The villages also depended on the farmers for their business, buying feed for their animals, selling their milk to the local cheese factory, buying hardware supplies, repairing farm equipment, and so on.

Beginning after World War II, from 1945 to 1960, agriculture experienced a dramatic revolution as electricity came to the countryside, tractors replaced horses, hybrid seeds became available, and much more. Farm numbers plummeted; the few farms left grew larger. By the 1970s, many of these small, rural villages had become mere shadows of what they once were, as Main Street businesses closed their doors, and the buildings stood vacant.

Farmers who had retired and moved to the villages made up much of the villages' population for several years, as they struggled to continue on. Some of them attracted small businesses and industries, which provided employment. Some built retirement housing, nursing homes, and clinics for their decidedly older population.

By the early 1960s, all of the one-room country schools had closed, and the few young people remaining in the community were bused to the consolidated schools in the villages. Some of the villages were close to lakes and other recreational opportunities (trout streams, for example) and attracted urban people to the area. These visitors supported the restaurants, craft shops, motels, and gasoline stations.

Then in 2020, COVID-19 appeared and swept across the country, indeed around the world. Retail businesses closed, which included many restaurants. Hospitals were filled with COVID cases. Thousands of people died. People were told to get vaccinated and stay home. And if they, of necessity, needed to leave home, they must wear a mask and keep six feet away from each other.

Many people were offered the opportunity to work from home, as a lot of businesses struggled to continue. Soon thousands of people were working from their homes, which for those in the big cities often meant a small, expensive apartment. As I mentioned, it quickly occurred to many that they could continue working at home, even after COVID subsided. Home could be almost anyplace that had a good broadband internet connection. People remembered rural communities that they had visited, or where they had relatives. They remembered what it was like when they vacationed in these communities. They enjoyed the quiet, the open spaces, and the recreational opportunities. Their children enjoyed these places as well. And some of them began moving to these rural communities, making them their new homes. The rural villages of the future, if they develop along the lines outlined in this book, could include the following categories of residents.

First, the "new farmers" that I described above, who farm in a variety of ways, ranging from following the tenets of organic farming and small acreage farming to doing more exotic types of agriculture, such as hydroponics and aquaculture, will help support these rural villages and cities. The rural villages and cities will benefit from these new farmers by having farm-to-table restaurants, perhaps fresh food from farmers' grocery cooperatives, and farmers' markets during the warmer months of the year.

A second group of residents in these future rural villages will include the work-at-home professionals, who have moved with their families from the big cities to the country. And not to forget the retired farmers who sold their farms and moved to town.

Recently retired urban people make up a third group of those seeking an alternative to urban living. They bring with them a disposable income and often new ideas for community improvement, as well as volunteering for various roles in community historical societies, libraries, and various cultural activities the community may sponsor. As these retired people grow older, they may challenge existing health-care systems in the community to help it respond to older persons' needs.

Following the suggestions offered earlier, these rural villages can thrive as viable alternatives to those living in major urban areas and the suburban areas that surround them. To make them attractive to these new residents, they must have broadband internet service, good medical service, excellent educational opportunities, cultural opportunities including good libraries and local historical societies, and above all, an attitude that accepts change as a necessity for the rural villages to survive.

Learning From Each Other

Finally, in addition to the need for a new agriculture and rural villages that keep up with the changes occurring in rural America, it is important for rural and urban America to work together as the country moves into the future. An important first step is for the two groups to get better acquainted with each other. County and state fairs are important ways to do this as the two groups come together to be entertained as well as learn from each other. Watching animals and farm produce being judged and awarded ribbons. Seeing 4-H and Future Farmers of America member exhibits, ranging from various farm animals to youth projects, provide an easy means for urban people to learn about rural life. Grandstand events, from 4-H dress reviews to tractor pulling contests to live entertainment also draw many people to a fair. Of course, the midway, with the Ferris wheel, merry-go-round, and food tents offering everything from cotton candy to corn dogs, brings loads of both urban and rural people to a fair.

Farmers' markets, large and small, where farmers bring their produce to sell, is another great way for rural and urban people to meet and learn from each other—as well as provide a market for fresh-grown produce.

Many rural communities sponsor festivals of various kinds during the summer months. My home community of Wild Rose, Wisconsin, each year offers Wild Rose Days: Party on the Pond. Activities include a parade down Main Street, fireworks, games, live entertainment, food, and much more. The event attracts hundreds of people each year.

A thriving, new agriculture along with vibrant rural villages and cities are essential for a thriving country. But changes must be made, some of them rather dramatic. My hope is that this book has offered some ideas for the changes that are necessary for both agriculture and rural villages and cities to thrive once more.

Acknowledgments

Many people helped me with this book. A huge thanks to the following people who read a draft of the manuscript and offered suggestions for improvement: Dennis Boyer, fellow emeritus, the Interactivity Foundation and director of policy projects on agriculture and rural life; Andy Lewis, professor emeritus, University of Wisconsin Extension; Jeff Wild, pastor and author; Philip Hasheider, farmer and author; Richard L. Cates, Jr., co-owner Cates Family Farm and senior lecturer emeritus, Department of Soil Science, University of Wisconsin–Madison; Katherine J. Cramer, professor, political science, University of Wisconsin–Madison; Paul Bodilly, retired middle school science teacher; and Jeffrey W. Apps, vice president—Private Client Services at SRS Capital Advisors (my son).

I especially want to thank Sam Scinta and Alison Auch of Fulcrum Publishing for reading an early draft of the book and offering several excellent ideas for giving a tighter focus to the work. As with all of my writing, my wife, Ruth, read and made comments as I wrote this book. Her help is more appreciated than she will ever know. Additionally, I want to thank my son, Steve, my daughter-in-law Natasha, and my daughter, Sue, who have supported my work on this book in an untold number of ways.

Notes

1. NCSL, "Challenges Facing Rural Communities," updated January 21, 2020, https://www.ncsl.org/agriculture-and-rural-development/challenges-facing-rural-communities#:~:text=Rural%20communities%20face%20challenges%20related,and%20environment%20and%20community%20preservation.
2. Trading Economics, "United States—Rural Population," https://tradingeconomics.com/united-states/rural-population-percent-of-total-population-wb-data.html.
3. United States Census Bureau, "2000 Census Urban and Rural Classification," https://www.census.gov/programs-surveys/geography/guidance/geo-areas/urban-rural/2000-urban-rural.html.
4. Ben Miller, "Nearly Half of U.S. Cities Have Fewer Than 1,000 Residents," Government Technology, December 3, 2018, https://www.govtech.com/data/

nearly-half-of-us-cities-have-fewer-than-1000-residents.html.

5. Delana Lefevers, “Most People Don’t Know These 15 Super Tiny Towns in Nebraska Exist,” Only in Your State, September 24, 2021, https://www.onlyinyourstate.com/nebraska/tiny-towns-ne/.
6. “10 of the Tiniest Towns in Iowa,” 98.1 KHAK, June 20, 2017, https://khak.com/10-of-the-tiniest-towns-in-iowa/.
7. Wisconsin Taxpayers Alliance and The League of Wisconsin Municipalities, “The 2017 State of Wisconsin’s Cities and Villages,” https://www.lwm-info.org/DocumentCenter/View/1605/2017-State-of-WI-Cities-and-Villages-report.
8. Jordan Institute for Families, “Facts about North Carolina’s Small Towns,” June 2007, https://practicenotes.org/vol12_no3/smalltown.htm.
9. Jerry Apps, *Simple Things: Lessons from the Family Farm* (Madison: Wisconsin Historical Society Press, 2018), 128–132.
10. Find My Past, “1830 U.S. Census Quick Facts,” https://www.findmypast.com/articles/world-records/full-list-of-united-states-records/census-land-and-substitutes/us-census-1830.

11. Jeff Hoyt, "1800–1990: Changes in Urban/Rural U.S. Population," Seniorliving.org, updated July 27, 2023, https://www.seniorliving.org/history/1800-1990-changes-urbanrural-us-population/; National Archives, "1940 Census FAQs," reviewed August 15, 2022, https://www.archives.gov/research/census/1940/faqs.
12. "Buck and Bright, Faithful Oxen, Helped Pioneers Carve Out Farms," *Antigo Daily Journal*, April 2, 1930.
13. Reuben Gold Thwaites, *Stories of the Badger State* (New York: American Book Company, 1900), 173.
14. Mark Wyman, *The Wisconsin Frontier* (Bloomington: Indiana University Press), 180–181.
15. National Archives, "The Homestead Act of 1862," reviewed June 2, 2021, https://www.archives.gov/education/lessons/homestead-act?_ga=2.86682264.1261758242.1697144649-1302466597.1697144567#background.
16. History, "Homestead Act," updated September 13, 2022, https://www.history.com/topics/american-civil-war/homestead-act.
17. History, "Dust Bowl," updated April 24, 2023, https://www.history.com/topics/great-depression/dust-bowl#.

18. Chautauqua, Boulder, CO, "The Chautauqua Movement," https://www.chautauqua.com/2021/chautauqua-movement-history/.
19. National Archives, "Records of the Rural Electrification Administration," https://www.archives.gov/research/guide-fed-records/groups/221.html.
20. Wisconsin Crop and Livestock Reporting Service, *Wisconsin Agriculture in Mid-Century* (Madison: Wisconsin State Department of Agriculture Bulletin No. 325, 1953), 22.
21. USDA, Economic Research Service, "Farming and Farm Income," updated August 31, 2023, https://www.ers.usda.gov/data-products/ag-and-food-statistics-charting-the-essentials/farming-and-farm-income/#:~:text=After%20peaking%20at%206.8%20million,and%20increased%20nonfarm%20employment%20opportunities.
22. Jerry Apps, *Wisconsin Agriculture: A History* (Madison: Wisconsin Historical Society Press, 2015), 215–219.
23. Iman Ghosh, "From Concrete Jungles to Crop Fields: This Is How America Uses Its Land," World Economic Forum and Visual Capitalist, February 11, 2020, https://www.weforum.org/agenda/2020/02/

land-use-america-agriculture-nature-urban.

24. Ghosh, "From Concrete Jungles to Crop Fields."
25. USDA, Economic Research Service, "Farming and Farm Income," updated August 31, 2023, https://www.ers.usda.gov/data-products/ag-and-food-statistics-charting-the-essentials/farming-and-farm-income/.
26. Olugbenga Ajilore and Caius Z. Willingham, "Redefining Rural America," Center for American Progress, July 17, 2019, https://www.americanprogress.org/article/redefining-rural-america/.
27. Andy Lewis, personal correspondence, August 11, 2022.
28. World Population Review, "Most Rural States," updated August 2023, https://worldpopulationreview.com/state-rankings/most-rural-states.
29. Anne N. Junod, Clare Salerno, and Corianne Payton Scally, "Debunking Three Myths about Rural America," Urban Institute, October 30, 2020, https://www.urban.org/urban-wire/debunking-three-myths-about-rural-america.
30. Lumenlearning.com, "14.4 Problems of Rural Life," https://courses.lumenlearning.com/suny-social problems/chapter/14-4-problems-of-rural-life/.

31. Elizabeth A. Dobis, Thomas P. Krumel, Jr., John Cromartie, Kelsey L. Conley, Austin Sanders, and Ruben Ortiz, "Rural America at a Glance," USDA, Economic Research Service, November 2021, https://www.ers.usda.gov/webdocs/publications/102576/eib-230.pdf?v=6750.4.
32. Walmart, "Walmart History," https://corporate.walmart.com/about/history.
33. ScrapeHero, "Number of Family Dollar Stores in the United States in 2023," September 12, 2023, https://www.scrapehero.com/location-reports/Family%20Dollar-USA/.
34. Annette Alvarez, "Native American Tribes and Economic Development," Urban Land, April 19, 2011, https://urbanland.uli.org/development-business/native-american-tribes-and-economic-development/.
35. Indigenous Corporate Training, Inc., "What Is the Seventh Generation Principle?" May 30, 2020, https://www.ictinc.ca/blog/seventh-generation-principle.
36. Winona LaDuke, *All Our Relations: Native Struggles for Land and Life* (Cambridge, MA: South End Press, 1999), 1, 2.
37. Robin Wall Kimmerer, *Braiding Sweetgrass: Indigenous Wisdom, Scientific Knowledge, and the Teachings*

of Plants (Minneapolis: Milkweed Editions, 2013), 37.

38. Aldo Leopold, *A Sand County Almanac* (New York: Oxford University Press, 1949), 210.
39. Leopold, *A Sand County Almanac*, 202–204.
40. NASA, Global Climate Change, "What Is Climate Change?" https://climate.nasa.gov/what-is-climate-change/.
41. USDA, "Climate Solutions," https://www.usda.gov/climate-solutions.
42. US EPA, "Global Greenhouse Gas Emissions Data," updated February 15, 2023, https://www.epa.gov/ghgemissions/global-greenhouse-gas-emissions-data.
43. Ryan Hobert and Christine Negra, "Climate Change and the Future of Food," United Nations Foundation, September 1, 2020, https://unfoundation.org/blog/post/climate-change-and-the-future-of-food/.
44. Jason Mark, "The New Abnormal," *Sierra*, Winter 2021.
45. Rebecca Tippett, "Agriculture and Food Statistics: USDA Charts the Essentials," Carolina Demography, May 11, 2015, https://carolinademography.cpc.unc.edu/2015/05/11/agriculture-and-food-statistics-usda-charts-the-essentials/.
46. Ajilore and Willingham, "Redefining Rural America."

47. Dee Laninga and Anna Straus, "Big Ag Mythbusters: Is Industrial Agriculture Really Inevitable?" Farm Action, October 12, 2021, https://farmaction.us/2021/10/12/big-ag-mythbusters-is-industrial-agriculture-really-inevitable/.
48. Kristina Kiki Hubbard, "The Sobering Details Behind the Latest Seed Monopoly Chart," Civil Eats, January 11, 2019, https://civileats.com/2019/01/11/the-sobering-details-behind-the-latest-seed-monopoly-chart/.
49. Austin Frerick, "To Revive Rural America, We Must Fix Our Broken Food System," The American Conservative, February 27, 2019, https://www.theamericanconservative.com/to-revive-rural-america-we-must-fix-our-broken-food-system/.
50. Laninga and Straus, "Big Ag Mythbusters."
51. Wendell Berry, "The Agrarian Standard," in *The Essential Agrarian Reader: The Future of Culture, Community, and the Land*, ed. Norman Wirzba (Lexington: University Press of Kentucky, 2003), 24–26.
52. Aldo Leopold, "The Land Ethic," in "A Sustainable Future," PBS Online: *Death of the Dream.*
53. Mark Shephard, *Restoration Agriculture: Real-World Permaculture for Farmers* (Greeley, CO: Acres U.S.A., 2013), 45.

54. USDA, Natural Resources Conservation Service, "Nutrient Management," https://www.nrcs.usda.gov/getting-assistance/other-topics/nutrient-management.
55. NBC News, "What's the Stench? A Pile of Cow Manure," January 28, 2005, https://www.nbcnews.com/id/wbna6879097.
56. Iowa Pork Producers Association, "2020 Pork Industry Facts," https://www.iowapork.org/newsroom/facts-about-iowa-pork-production.
57. Jamie Konopacky and Soren Rundquist, "EWG Study and Mapping Show Large CAFOs in Iowa Up Fivefold Since 1990," EWG, January 21, 2020, https://www.ewg.org/interactive-maps/2020-iowa-cafos/.
58. Wisconsin Department of Natural Resources, "CAFO WPDES Permits Applications and NMPS in Process," https://dnr.wisconsin.gov/topic/CAFO/RecentPermits.html; USDA Natural Agricultural Statistics Service, "2011 Wisconsin Agricultural Statistics," https://www.nass.usda.gov/Statistics_by_State/Wisconsin/Publications/Annual_Statistical_Bulletin/bulletin2011_web.pdf.
59. A Greener World, "Pollution," https://agreenerworld.org/challenges-and-opportunities/environmental-pollution/.

60. Will Cushman, "CAFO Oversight in Wisconsin and Who Pays for It?" Wisconsin State Farmer, June 19, 2019, https://www.wisfarmer.com/story/news/2019/06/19/what-cafo-oversight-and-who-pays-wisconsin/1452486001/#.
61. Scott Gordon, "What Manure Digesters Can and Can't Do," WisCONTEXT, https://wiscontext.org/what-manure-digesters-can-and-cant-do.
62. Sara Popescu Slavikova, "Advantages and Disadvantages of Monoculture Farming," Greentumble, June 16, 2019, https://greentumble.com/advantages-and-disadvantages-of-monoculture-farming.
63. Slavikova, "Advantages and Disadvantages of Monoculture Farming."
64. Eric Hamilton, "Midwest Bumble Bees Declined with More Farmed Land, Less Diverse Crops Since 1870," University of Wisconsin–Madison, June 22, 2021, https://news.wisc.edu/midwest-bumble-bees-declined-with-more-farmed-land-less-diverse-crops-since-1870/#:~:text=Explore%20Topics-,Midwest%20bumble%20bees%20declined%20with%20more%20farmed,less%20diverse%20crops%20since%201870&text=As%20farmers%20cultivated%20more%20land,became%20rarer%20in%20Midwestern%20states.

65. P. Byrne, "Genetically Modified (GM) Crops: Techniques and Applications," Colorado State University Extension, https://extension.colostate.edu/docs/pubs/crops/00710.pdf.
66. Sheldon Krimsky and Jeremy Gruber, eds., *The GMO Deception* (New York: Skyhorse Publishing, 2014), xxi–xxv.
67. Laninga and Straus, "Big Ag Mythbusters."
68. "Big Ag Mythbusters: Do We Really Need Industrial Agriculture to Feed the World?" Farm Action, August 27, 2021, https://farmaction.us/2021/08/27/do-we-really-need-industrial-agriculture-to-feed-the-world/.
69. "Big Ag Mythbusters."
70. "Farm Bill," Food, Agriculture, Conservation and Trade Act of 1990 (FACTA), Public Law 101-624, Title VI, Subtitle A., Section 1603.
71. Union of Concerned Scientists, "What Is Sustainable Agriculture?" updated March 15, 2022, https://www.ucsusa.org/resources/what-sustainable-agriculture.
72. Superior Fresh, https://www.superiorfresh.com/our-farm.
73. Raoul Adamchak, "Organic Farming," Britannica, updated August 25, 2023, https://www.britannica.com/topic/organic-farming.

74. USDA Organic, "Guidelines for Organic Crop Certification," https://www.ams.usda.gov/sites/default/files/media/Crop%20-%20Guidelines.pdf.
75. Youmatter, "Organic Farming: Definition, Standards, Benefits," updated February 5, 2019, https://youmatter.world/en/definition/organic-farming-definition-standards-benefits/.
76. Jerry Apps, *Settlers Valley* (Madison: University of Wisconsin Press, 2021).
77. The Villages Grown, https://thevillagesgrown.com.
78. Eden Green, https://www.edengreen.com.
79. Ron Dongoski, "Next Up: Mark Oshima—Farming Up, EY Americas," July 15, 2020, https://www.ey.com/en_us/purpose/next-up-mark-oshima-farming-up.
80. Tori Avey, "From Pythagorean to Pescatarian: The Evolution of Vegetarianism," The History Kitchen, PBS, January 28, 2014, https://www.pbs.org/food/the-history-kitchen/evolution-vegetarianism/.
81. Anna Keeve, "Plant-Based Menu Items Are Infiltrating Fast Foods—and Meat-Eaters Are All Over Them," *Insider*, September 27, 2021, https://www.businessinsider.com/meat-eaters-opting-for-plant-based-vegan-fast-food-2021-9.
82. Keeve, "Plant-Based Menu Items Are Infiltrating Fast Foods."

83. Nate Crosser, "New GFI State of the Industry Reports Show Alternative Proteins Are Poised to Flourish Post-Covid-19," Good Food Institute, May 13, 2020, https://gfi.org/blog/state-of-the-industry-2020/.
84. M. Shahbandeh, "Per Capita Consumption of Dairy Products in the United States from 2000 to 2021," Statista, November 16, 2022, https://www.statista.com/statistics/183717/per-capita-consumption-of-dairy-products-in-the-us-since-2000/.
85. Susan S. Lang, "Lactose Intolerance Seems Linked to Ancestral Struggles with Harsh Climates and Cow Diseases, Cornell Study Finds," Cornell Chronicle, June 1, 2005, https://news.cornell.edu/stories/2005/06/lactose-intolerance-linked-ancestral-struggles-climate-diseases.
86. Lacey Muinos, "The 9 Best Vegan Cheeses of 2023," The Spruce Eats, updated October 6, 2023, https://www.thespruceeats.com/best-dairy-free-cheeses-1001581.
87. Steffani Sassos, "All of the Nutritional Facts and Health Benefits of Goat Milk," *Good Housekeeping*, April 14, 2020, https://www.goodhousekeeping.com/health/diet-nutrition/a32068757/goat-milk-health-benefits/.

88. Perfect Day, https://perfectday.com.
89. Alexandra Wilson, "Got Milk? This $40M Start-Up Is Creating Cow-Free Dairy Products That Actually Taste Like the Real Thing," *Forbes*, January 9, 2019.
90. Chris Harris, "After Generations of Disinvestment, Rural America Might Be the Most Innovative Place in the U.S.," Ewing Marion Kauffman Foundation, December 14, 2020, https://www.kauffman.org/currents/rural-america-most-innovative-place-in-united-states/.
91. Harris, "After Generations of Disinvestment."
92. Jack Healy, "Farm Country Feeds America. But Just Try Buying Groceries There," *New York Times*, November 5, 2019, https://www.nytimes.com/2019/11/05/us/rural-farm-market.html.
93. Viroqua, Wisconsin, Driftless Wisconsin, https://driftlesswisconsin.com/viroqua/.
94. Jerry Apps, *Whispers and Shadows: A Naturalist's Memoir* (Madison: Wisconsin Historical Society Press, 2015), 127.
95. Jerry Apps, *Old Farm: A History* (Madison: Wisconsin Historical Society Press, 2013).
96. Kimmerer, Braiding Sweetgrass, 36.

97. John Ikerd, *A Return to Common Sense* (Philadelphia: Edwards Publishing, 2007), v.

98. Norman Wirzba, *The Paradise of God: Renewing Religion in an Ecological Age* (New York, Oxford University Press, 2003), 2.

99. Author's interview with Kathy Cramer, October 26, 2022.

100. Jerry Apps, *Telling Your Story: Preserving Your History through Storytelling* (Golden, CO: Fulcrum Publishing, 2016).

101. Michael Ekpenyong Asuquo et al., "What Is Indigenous Knowledge?" IGI Global, InfoScipedia, 2023, https://www.igi-global.com/dictionary/communities-of-practice-and-indigenous-knowledge/56613.

102. E. F. Schumaker, *Small Is Beautiful: Economics as if People Mattered* (New York: Harper Perennial reprint edition, 2010).

103. Yasmin Anwar, "Nature Is Proving to Be Awesome Medicine for PTSD," *Berkeley News*, July 12, 2018, https://news.berkeley.edu/2018/07/12/awe-nature-ptsd/#:~:text=The%20awe%20we%20feel%20in,during%20white%2Dwater%20rafting%20trip.

104. Apps, *Settlers Valley*.

105. Cramer interview.

106. Cyndie Shearing, "Women Count in Agriculture," FB.org, https://www.fb.org/focus-on-agriculture/women-count-in-agriculture#:~:text=The%20Agriculture%20Department%27s%20just%2Dun-veiled,acres%20and%20value%20of%20production.

Additional Reading

Apps, Jerry. *Old Farm: A History*. Madison: Wisconsin Historical Society Press, 2013.

Apps, Jerry. *Settlers Valley*. Madison: University of Wisconsin Press, 2021.

Apps, Jerry. *Simple Things*. Madison: Wisconsin Historical Society Press, 2018.

Apps, Jerry. *Telling Your Story: Preserving Your History through Storytelling*. Golden, CO: Fulcrum Publishing, 2016.

Apps, Jerry. *The Land Still Lives*. Madison: Wisconsin Historical Society Press, 1970, 2019.

Apps, Jerry. *Whispers and Shadows: A Naturalist's Memoir*. Madison: Wisconsin Historical Society Press, 2015.

Apps, Jerry. *Wisconsin Agriculture: A History*. Madison: The Wisconsin Historical Society Press, 2015.

Apps, Jerry, and Natasha Kassulke. *Planting an Idea: Critical and Creative Thinking About*

Environmental Problems. Wheat Ridge, CO: Fulcrum Publishing, 2023.

Bass, Diana Butler. *Grounded: Finding God in the World*. New York: HarperOne, 2015.

Carson, Rachel. *Silent Spring*. New York: Houghton Mifflin, 1962.

Cramer, Katherine J. *The Politics of Resentment: Rural Consciousness in Wisconsin and the Rise of Scott Walker*. Chicago: The University of Chicago Press, 2016.

Danborn, David B. *Born in the Country: A History of Rural America*. Baltimore: Johns Hopkins University Press, 1995, 2005.

Gore, Al. *Earth in the Balance: Ecology and the Human Spirit*. New York: Plume, 1993.

Hanson, Victor Davis. *Fields Without Dreams: Defending the Agrarian Idea*. New York: The Free Press, 1996.

Hart, John Fraser. *The Land That Feeds Us: The Story of American Farming*. New York: W. W. Norton, 1991.

Ikerd, John. *A Return to Common Sense*. Philadelphia: Edwards Publishing, 2007.

Kimbrell, Andrew, ed. *The Fatal Harvest Reader: The*

Tragedy of Industrial Agriculture. Sausalito, CA: Foundation for Deep Ecology, 2002.

Kimmerer, Robin Wall. *Braiding Sweetgrass: Indigenous Wisdom, Scientific Knowledge, and the Teachings of Plants*, Minneapolis, MN: Milkweed Editions, 2013.

Leopold, Aldo. *A Sand County Almanac*. New York: Oxford University Press, 1949, 1968.

Mock, Sarah K. *Farm (And Other F Words): The Rise and Fall of the Small Family Farm*. Potomac, MD: New Degree Press, 2021.

Pollan, Michael. *Omnivore's Dilemma: A Natural History of Four Meals*. New York: Penguin Books, 2006.

Pyle, George. *Raising Less Corn, More Hell: The Case for the Independent Farm and Against Industrial Food*. New York: Public Affairs, 2005.

Schumacher, E. F. *Small Is Beautiful: Economics as if People Mattered*. New York: Vintage, 1993.

Shepard, Mark. *Restoration Agriculture: Real-World Permaculture for Farmers*. Greeley, CO: Acres U.S.A., 2013.

Wheeler, Stephen M., and Christina D. Rosan. *Reimagining Sustainable Cities: Strategies for*

Designing Greener, Healthier, More Equitable Communities. Oakland, CA: University of California Press, 2021.

Wirzba, Norman, ed. *The Essential Agrarian Reader: The Future of Culture, Community, and the Land*. Lexington: University Press of Kentucky, 2003.

Wirzba, Norman. *The Paradise of God: Renewing Religion in an Ecological Age*. New York: Oxford University Press, 2003.

Other Books in the Speaker's Corner Series

On Censorship: A Public Librarian Examines Cancel Culture in the US, by James LaRue

---

This book is essential reading for all those who believe in free expression, who support libraries, and who cherish the central freedoms of American democracy.

On Digital Advocacy: Saving the Planet While Preserving Our Humanity, by Katie Boué

The guidebook for saving the planet while preserving and protecting your human spirit.

On the Gaze: Dubai and Its New Cosmopolitanisms, by Adrianne Kalfopoulou

---

An immersive experience of Dubai complete with vivid portraits, elegant prose, and historical context.

On Indigenuity: Learning the Lessons of Mother Earth, by Daniel R. Wildcat

---

Our earth needs care, and this book teaches us how to give that care and respect the way Indigenous people have for thousands of years.

About the Author

PHOTO BY STEVE APPS

Jerry Apps is a former county extension agent and is now professor emeritus at the University of Wisconsin–Madison, where he taught for thirty years. Today he works as a rural historian and full-time writer and is the author of many books on rural history, country life, and the environment. He has created seven, hour-long documentaries with PBS Wisconsin, has won several awards for his writing, and won a regional Emmy Award for the TV program *A Farm Winter*. Jerry and his wife, Ruth, have three children, seven grandchildren, and three great-grandsons. They divide their time between their home in Madison and their farm, Roshara, in Waushara County.